Astronomy in India: A Historical Perspective

Edited by

Thanu Padmanabhan

Indian National Science Academy
Platinum jubilee special volume

Indian National Science Academy

Springer

Editor

Prof. Thanu Padmanabhan
IUCAA
Post Bag 4,
Pune University,
Pune - 411007

ISBN 978-81-8489-998-6 (eBook)

Padmanabhan: Astronomy in India

Softcover reprint of the hardcover 1st edition 2014

First Published 2010

ISBN 978-81-322-2844-8
DOI 10.1007/978-81-8489-998-6

This edition is published by Springer (India) Private Limited, (a part of Springer Science+Business Media), Registered Office: 3rd Floor, Gandharva Mahavidyalaya, 212, Deen Dayal Upadhyaya Marg, New Delhi – 110 002, India.

Foreword

The Indian National Science Academy (INSA) was established in January 1935 with the objective of promoting science in India and harnessing scientific knowledge for the cause of humanity and national welfare. The Academy has laid special emphasis on the recognition and promotion of scientific excellence. It has also concerned itself with many societal issues. It has a vigorous international programme. For instance, the Academy is the adhering body in India of the International Council for Science (ICSU) on behalf of the Government of India. Furthermore, it has active bilateral exchange programmes with several sister academies around the world. The Academy has also been involved in the publication of journals, books, special volumes, biographical memoirs, etc.

The Platinum Jubilee events of the Academy were inaugurated by the Prime Minister of India on January 10, 2009 at New Delhi. In addition to new initiatives such as those on Science Policy Study Cell and INSA Archives, several scientific programmes are being organised across the country during the Platinum Jubilee Year. Many special publications on a variety of subjects are also being brought out during the year.

The year 2009 has been declared as the International Year of Astronomy by the United Nations to celebrate the 400th anniversary of the use of telescope by Galileo to study heavenly bodies. Thus, happily, the Platinum Jubilee Year of INSA and the International Year of Astronomy overlap. Therefore, it is appropriate for INSA to bring out a volume on astronomy, both ancient and modern. This volume containing authoritative reviews, edited by Professor T. Padmanabhan, have contributions by leading scientists of the country. Of the six chapters, two deal with ancient traditions of astronomy in India including the development of calculus while the remaining four discuss the contemporary scenario. I hope this volume will both inform and entertain the interested reader.

M. Vijayan
President
Indian National Science Academy

Preface

India has a strong and ancient tradition of astronomy which seamlessly merges with the current activities in astronomy and astrophysics in the country. While the younger generation of astronomers and students are reasonably familiar with the current facilities and the astronomical research, one often notices that they do not have an equally good knowledge of the history of Indian astronomy. This particular volume, brought out as a part of the Platinum Jubilee celebrations of INSA, concentrates on selected aspects of historical development of astronomy in this country.

It is, of course, impossible to do justice in a single volume to this rich and varied topic and hence I needed to make some choices (which obviously will not make everyone happy!). I have tried to combine the ancient with the modern by choosing some representative topics from both eras.

As a result you find two chapters by Balachandra Rao and M.S. Sriram dealing with the development of ancient Indian astronomy and the development of calculus in the form of the ancient Kerala text *Yuktibhasa*. I am sure some of the details in these two chapters will come as a surprise to many young astronomers. The other four chapters are more contemporary. Siraj Hasan, Jayant Narlikar, B.V. Sreekantan and G. Swarup have highlighted the historical development of Optical Astronomy, Relativistic Astrophysics, Space Astronomy and Radio Astronomy, respectively. Here the choice I have made covers one area of theory and three bands of the electromagnetic spectrum in which serious research is going on in this country.

I thank all the contributors for their contributions as well as Mrs. Manjiri Mahabal for formatting and latexing the volume. The further processing of the manuscript was done by Professor Alok K. Gupta, Dr. Mrigank M. Dwivedi, Mr. Soumitra Dasgupta and Mr. Abhay P. Singh, I am grateful to them for their help.

T. Padmanabhan
Editor

Editorial Board

Contents

Chapter 1
Classical Astronomy in India – An Overview

S. Balachandra Rao

yatha sikha mayuranam naganam manayo yatha |
tadvad vedanga sastranam jyotisam (ganitam) murdhani sthitam ||

"Like the crests on the heads of peacocks, like the gems on the hoods of the cobras, Astronomy (Mathematics) is at the top of the *Vedanga sastras* – the auxiliary branches of the Vedic knowledge."

Vedganga Jyotisam, R-VJ, 35; *Y-VJ,* 4 (Sastry 1985)

1.1 Introduction

In the present chapter, an attempt is made to survey, though briefly, the development of and contribution to astronomy through the ages – from the Vedic period up to the colonial period. Some significant contributions are highlighted.

One remarkable feature of our Indian mathematicians and astronomers is their undaunted attack on tradition whenever they found fault with it. In fact, Bhaskara II (b. 1114 AD) assets, "*It is necessary to speak out the truth before those who have implicit faith in tradition. It will be impossible to believe in whatever is said earlier unless every erroneous statement is criticized and condemned.*"

While astronomy was used primarily for fixing auspicious time for religious rites, mathematics mostly served as the hand-maid of astronomy. An over-emphasis on religious utilitarianism of mathematics and astronomy seems to have stunted the growth these two fields now and then. The nonchalance with which the splendid achievements of Greek geometry – accompanied with the rich tradition of deductive mathematical logic – were ignored (in 3rd–4th century BC) while the pseudo-science of Greek and Babylonian astrology (*hora sastra*) was received, hugging with open arms, is indeed the worst of these! But then, there are better informed scientists among us who assert that there was *no need* for ancient Indians to borrow Euclid's geometry since we had even richer geometry in the more ancient *Sulvasutras* texts like those of *Baudhayana* and *Apastamba*.

1.2 Astronomy of the Vedas

The verse cited at the beginning of this chapter shows the supreme importance given to astronomy (and mathematics) among the branches of knowledge ever since the Vedic times. Like many other branches of knowledge, even the beginnings of the science of astronomy in India have to be traced back to the Vedas. In the Vedic literature, *Jyotisa* is one of the six auxiliaries (*sadangas*) of the corpus of Vedic knowledge. The six *Vedangas* are (Rao 2005):

(1) *Siksa* (phonetics); (2) *Vykarana* (grammar); (3) *Chandas* (metrics); (4) *Nirukta* (etymology); (5) *Jyotisa* (astronomy); and (6) *Kalpa* (rituals).

Hon. Director, Gandhi Centre of Science and Human Values, 43/1, Race Course Road, Bangalore - 560 001.
e-mail: balachandra_rao@rediffmail.com

It is important to note that although in modern common parlance the word *Jyotisa* is used to mean predictive astrology, in the ancient literature *Jyotisa* meant all aspects of astronomy. Of course, mathematics was regarded as a part of *Jyotisa*. The *Vedanga Jyotisa* is the earliest Indian astronomical text available.

Even during the time of the early *mandalas* of the *Rigveda* the astronomical knowledge necessary for the day-to-day life of the people was acquired. The Vedic people were conversant with the knowledge required for their religious activities, like the time (and periodicity) of the full and the new moons, the last disappearance of the moon and its first appearance, etc. This type of information was necessary for the monthly rites like *darsapurnamasa* and seasonal rites like *caturmasya*.

1.2.1 Mention of Naksatras

Since the sidereal period of the moon was known for long as around 27.3 days, the presence of the moon amidst a star group each day of this period was observed over long periods of time. Each such star group of the moon's sojourn, approximately over a day of the moon's sidereal period, was identified by the most brilliant star in that group. Those identifying stars (27 or 28) are called *yogataras* and named as *Asvini, Bharani*, ..., *Revati*. However, during the Vedic period, the vernal equinox did not occur in the *Asvini* asterism. The *Rigveda Samhita* mentions only a few of the 27 *naksatras*. Those asterisms are *Tisya (Pusya), Citra, Revati, Agha (Magha), Arjuni (Phalguni)* (Sastry 1985; Dikshit 1969).

The *Taittiriya Samhita* gives the list of all the 28 *naksatras* including *Abhijit*. The *Satapatha Brahmana* mentions that the (main) *nakasatras* are 27 and the "*upanaksatras*" (subsidiary stars) are also 27. The *Rigveda Samhita*, for example mentions two *naksatras* viz. *Magha* and *Phalguni* in succession:

suryaya vahatuh pragat savitayam avasrjat |
aghasu hanyante gavorjunyoh paryuhyate ||

"The (dowry) of cows which was given by *Savita* (sun) had already gone ahead of *Surya* (sun's daughter). They drive the cows on the (day of) *Agha* (Magha) *naksatra*. The (daughter) was carried away on the (day) of *Arjuni* (Phalguni) star."

RV, 10, 85, 13

1.2.2 Mention of Planets

Among the planets it is *Brhaspati* (*Guru*, Jupiter) that finds explicit mention in the *Rigveda*:

brhaspatih prathamam jayamano maho jyotisah parame vyoman |
saptasyastu vijato ravera saptarasamiradhamat tamamsi ||

"*Brhaspati*, when being born in the highest heaven of supreme light, seven-mouthed, multiform (combined) with sound and seven-rayed, has subdued the darkness."

RV, 4, 50, 4

It is also argued that *Vena* mentioned in *Rigveda* (*RV*, 10, 123, 5) refers to *Sukra* (Venus).

The *naksatra* system consisting of 27 *naksatras* (or 28 including *Abhijit*) was evolved long back and was used to indicate days. It is pointed out that *Agrahayana*, an old name for the *Mrgasira naksatra*, meaning "beginning of the year" suggests that the sun used to be in that asterism at the vernal equinox. This corresponds to the period of around 4000 BC.

The *Rohini* legends in the *Rigveda* point to a time in the late *Rigveda* period when the vernal equinox shifted to the *Rohini* asterism (from *Mrgasira*). The later sacrificial session called *Gavamayana* was especially designed for the daily observation of the movements of the sun and of the disappearance of the moon. This must have given the priests and their advisors sufficient knowledge of a special kind, even like the "saros" cycle for predicting the eclipses. There is evidence, in the *Rigveda* that this specialised knowledge about the eclipses was possessed by the priests of the *Atri* family.

During the *Yajurveda* period, it was known that the solar year has 365 days and a fraction more. In the *Taittiriya Samhita* it is mentioned that the extra 11 days over the 12 lunar months (totalling to 354 days), complete the six *rtus* by the performance of the *ekadasaratra*, i.e. 11 nights sacrifice. Again, the same *Samhita* says that 5 days more were required over the *savana* year of 360 days to complete the seasons adding specifically that "4 days are too short and 6 days too long."

The Vedic astronomers evolved a system of 4 years' *yuga*. The names of the 4 years of a *yuga* are: (i) *Samvatsara,* (ii) *Parivatsara*, (iii) *Idavatsara* and (iv) *Anuvatsara* .

This period of a *yuga* (of 4 years) was used to reckon time as can be seen from the statements like, "*Dirghatamas*, son of *Mamata*, became old even in his 10th *yuga*," i.e. between the age of 45 and 50 years (*RV* 1.158.6). The two intercalary months, *Amhaspati* and *Samasarpa*, to complete the *yuga* (of 4 years) were known (*RV* 1.25.8).

In the *Yajurveda*, a year comprising 12 solar months and 6 *rtus* (seasons) was recognised. The grouping of the six *rtus* and the 12 months in the Vedic nomenclature is given in Table 1.1.

Table 1.1 *Rtus* and months in Vedic nomenclature

	Seasons	Months
1.	*Vasanta rtu*	*Madhu* and *Madhava*
2.	*Grisma rtu*	*Sukra* and *Suci*
3.	*Varsa rtu*	*Nabha* and *Nabhasya*
4.	*Sarad rtu*	*Isa* and *Urja*
5.	*Hemanta rtu*	*Saha* and *Sahasya*
6.	*Sisira rtu*	*Tapa* and *Tapasya*

The sacrificial year commenced with *Vasanta rtu*. The Vedic astronomers had also noted that the shortest day was at the winter solstice when the seasonal year *Sisira* began with *Uttarayana* and rose to a maximum at the summer solstice.

1.3 *Vedanga Jyotisa* (Sastry 1985)

The purpose of the *Vedanga Jyotisa* was mainly to fix suitable times for performing the different sacrifices. The text is found in two recensions – *Rigveda Jyotisa* and *Yajurveda Jyotisa*. Though the contents of both recensions are the same, they differ in the number of verses contained in them. While the *Rigvedic* version contains only 36 verses, the *Yajurvedic* version contains 44 verses. This difference in the number of verses is perhaps due to the addition of explanatory verses by the *adhvaryu* priests with whom it was in use.

In one of the verses, it is said, "I shall write on the lore of time, as enunciated by sage Lagadha." Therefore, the knowledge *Vedanga Jyotisa* is attributed to Lagadha. But the composition of the text could be of a much later period.

According to the text, at the time of its composition, the winter solstice was at the beginning of the constellation *Sravistha* (Dhanistha, Delphini) and the summer solstice was in the middle of the *Aslesa* constellation. Since Varahamihira (505 AD) stated that in his own time the summer solstice was at the end of three quarters of *Punarvasu* and the winter solstice at the end of the first quarter of *Uttarasadha*, there had been a precession of the equinoxes (and solstices) by one and three quarters of a *naksatra*, i.e. about $23°20'$. Since the rate of precession is about a degree in 72 years, the time interval for precession of $23°20'$ is about $72 \times 23°20'$, i.e. 1,680 years prior to Varahamihira's time. This takes us back to around 1175 BC. Generally, the accepted period of *Vedanga Jyotisa* is between 12th and 14th centuries BC. However, according to some scholars the style and language of the text suggest that its actual composition must have taken place a few centuries later.

The *Vedanga Jyotisa* belongs to the last part of the Vedic age. The text proper can be considered as the record of the essentials of astronomical knowledge needed for the day-to-day life of the people of those times. The *Vedanga Jyotisa* is the culmination of the knowledge of astronomy developed and accumulated over thousands of years of the Vedic period up to 1400 BC.

In the *Vedanga Jyotisa*, a yuga of 5 solar years consists of 67 lunar sidereal cycles, 1,786 days, 1,835 sidereal days, 62 synodic months, 1,860 *tithis*, 135 solar *naksatras*, 1,809 lunar *naksatras* and 1,768 risings of the moon. It also mentions that there are 10 *ayanas* and *visuvas* and 30 *rtus* in a *yuga*. The practical way of measuring time is mentioned as the time taken by a specified quantity of water to flow through the orifice of a specified clepsydra (water-clock) as one *nadika*, i.e. 1/60 part of a day.

One can find in the *Vedgana Jyotisa* very useful presentation of the various calendrical items prevalent during those times like:

1. the solstices
2. increase and decrease of the durations of days and nights in the *ayanas*
3. the solstitial *tithis*
4. the seasons
5. omission of *tithis*
6. table of *parvas*
7. *yogas* (which developed later as one of the five limbs of fullfledged *pancanga*)
8. finding the *parva naksatras* and the *parva tithis*
9. the *visuvas* (equinoxes)
10. the solar and other types of years
11. the revolutions of the sun and moon (as seen from the earth)
12. the times of the sun's and the moon's transit through a *naksatra*
13. the *adhikamasa* (intercalary month)
14. the measures of the longest day and the shortest night, etc.

The *Vedanga Jyotisa* mentions that the durations of the longest and the shortest days on the two solstitial days are of ratio 3 : 2, i.e. 36 and 24 *ghatikas* (or *nadikas*) which correspond to 4^h24^m and 9^h36^m respectively. This means the *dinardhas,* i.e. the lengths of half-days come to be 7^h12^m and 4^h48^m respectively. It is calculated that around 1400 BC, the sun's maximum declination δ used to be about $23°53'$. The current value of δ is about $23°27'$. However, our ancient Indian astronomers took it as 24°. Now, the latitude ϕ of a place can be found using the formula:

$$sin(\text{ascensional difference}) = tan\phi tan\delta \quad (1.1)$$

where δ is the declination of the sun. The correction due to ascensional difference in this case is 1^h12^m, i.e. in angular measure, $1\text{h}12\text{m} \times 15 = 18°$. Now, using the above formula, we get the latitude of the place, $\phi = 35°$ approximately. Therefore, the place of composition of the *Vedanga Jyotisa* appears to be in the region of Kashmir.

1.4 *Siddhantas*

The astronomical computations described in the *Vedanga Jyotisa* were in practical use for a very long time. Around the beginning of the Christian era, say a century on either side of it, a new class of Indian astronomical literature emerged. The texts representing this development are called *siddhantas*. The word "*siddhanta*" has the connotation of an established theory. These *siddhanta* texts contain much more material and topics than the *Vedanga Jyotisa* (Rao 2005). Along with the *naksatra* system, the 12 signs (*rasis* of the zodiac) viz. *Mesa, Vrsabha,* etc. were introduced. A precise value for the length of the solar year was adopted. Computations of the motions of the planets, the solar and lunar eclipses, ideas of parallax, determination of mean and true positions of planets and a few more topics formed the common contents of the *siddhantic* texts.

A very significant aspect of that period, in the history of Indian astronomy, was the remarkable development of newer mathematical methods which greatly promoted mathematical astronomy. For example, Aryabhata I's general solution for the first order indeterminate equation (*Kuttaka*) greatly helped in obtaining the *bhaganas* of heavenly bodies. Needless to say, the unique advantage of the famous Hindu invention of decimal numerals – adopted world over now – made computations with the huge numbers very handy and even enjoyable to the ancient Indian astronomers. According to the Indian tradition, there were principally 18 *siddhantas: Surya, Paitamaha, Vyasa, Vasistha, Atri, Parasara, Kasyapa, Narada, Gargya, Marici, Manu, Angira, Lomasa* (Romaka), *Paulisa, Cyavana, Yavana, Bhrgu* and *Saunaka*. However, among these only five *siddhantas* were extant during the time of Varahamihira (505 AD) viz. *Saura* (or *Surya*), *Paitamaha* (or *Brahma*), *Vasistha, Romaka* and *Paulisa*. These five *siddhantas* were ably collected together by Varahamihira and preserved for the posterity as his *Pancasiddhantika* (Sastry 1993).

1.4.1 Famous Astronomers and Their Works

Some of the famous Indian astronomers and their major astronomical works are listed in Table 1.2. The dates in brackets refer either to the approximate dates of composition of the works or of authors' birth (preceded by b.).

Table 1.2 Famous Indian astronomers and their major astronomical works

	Author	Works
1.	Aryabhata I (499 AD)	*Aryabhatiyam* (Shukla and Sharma 1976) *Aryasiddhanta* (not traced)
2.	Varahamihira (505 AD)	*Pancasiddhantika* (Sastry 1993), *Brahatsamhita* (Bhat and Banarasidass 1981)
3.	Bhaskara I (c. 600 AD)	*Bhasya* on *Aryabhatiyam* (Shukla 1976), *Mahabhaskariyam* (Shukla 1960), *Laghubhaskariyam* (Shukla 1960)
4.	Brahmagupta (628 AD)	*Brahmasphutasiddhanta* (Sharma 1966), Khandakhadyaka (Sengupta 1934; Chatterjee 1970)
5.	Vatesvara (880 AD)	*Vatesvarasiddhanta* (Shukla 1986)
6.	Manjula (932 AD)	*Laghumanasam* (Shukla 1990)
7.	Aryabhata II (c. 950 AD)	*Mahasiddhanta* (Dvivedi 1910)
8.	Bhaskara II (b. 1114 AD)	*Siddhantasiromani* (Dvivedi 1929; Arkasomayaji 1980) *Karanakutuhalam* (Rao and Uma 2007–08)
9.	Paramesvara (c. 1400 AD)	*Drgganitam* (Sarma 1963), *Suryasiddhanta* (Shukla 1957), *Vivaranam, Bhatadipika* (Kern 1990)
10.	Nilakantha Somayaji (c. 1500 AD)	*Tantrasangraha* (Sarma 1977), *Aryabhatiyabhasya* (Sastri 1977)
11.	Ganesa Daivajna (1520 AD)	*Grahalaghavam* (Rao and Uma 2006), *Titihicintamani* (Apte 2002)
12.	Jyestdeva (c. 1540 AD)	*Yuktibhasa* (Sarma et al. 2008)
13.	Candrasekhara Samanta (b.1835 AD)	*Siddhantadarpanah* (Upadhyaya 1996, 1998)
14.	Sankara Varman (19th century)	*Sadratnamala* (Sarma 2001)
15.	Venkatesa Ketkar (1898 AD)	*Grahaganitam* (Ketkar 2006), *Jyotirganitam* (Ketkar 2008)

Note: The dates of some of the above authors are debated.

1.4.2 Aryabhata I (b. 476 AD)

Aryabhata I, different from his namesake of the 10th century, was born in 476 AD and composed his very famous work, *Aryabhatiyam* when he was 23 year old. He mentions in his monumental text that he sets forth to present the knowledge honoured at Kusumapura, identified with modern Patna in Bihar.

The *Aryabhatiyam* consists of four parts (*padas*): *Gitika*, *Ganita, Kalakriya* and *Gola*. The first part contains 13 verses and the remaining three parts, forming the main body of the texts, contain a total of 108 verses.

In the *Gitikapada*, we are introduced to:

1. the large units of time viz. *Kalpa, Manvantara* and *Yuga* (different from that of the *Vedanga Jyotisa*);
2. circular units of arc viz. degrees and minutes; and
3. linear units viz. *yojana, hasta* and *angula*

The numbers of revolutions of planets in a *maha-yuga* of 43,20,000 years are given in the *Gitikapada*. Further, the positions of the planets, their apogees (or aphelia) and nodes are also given. Besides these, the diameters of the planets, the inclinations of the orbital planes of the planets with the ecliptic and the peripheries of the epicycles of the different planets are also included. The topic of great mathematical importance, in this part, is the construction of the tables of *Jya* the trigonometric function "sine" (in just one *sloka*). It is significant that so much of information is packed, as if in a concentrated capsule form, in just 10 verses. It has to be noted that $Jya\ (\theta) = R\ sin\ (\theta)$ where R is an arbitrary constant. The second part of the *Aryabhatiyam*, the *Ganitapada* contains 33 stanzas essentially dealing with mathematics. This part lays the strong base for all the remarkable future contribution of nearly thirteen centuries to mathematics in India. In fact, the most significant contribution of Aryabhata, in the history of world mathematics, is his method of general solution of a *first order indeterminate equations* (called *Kuttaka*): to find solutions of $ax + by = c$, in integers (where a and b are given integers).

The *Kalakriyapada*, the third part of the *Aryabhatiyam* contains 25 verses explaining the various units of time and the method of determination of positions of planets for a given day. Calculations concerning the *adhikamasa* (intercalary month), *ksayatithis*, angular speeds of planetary motions (in terms of revolutions), the concept of weekdays are all included in this part of the text.

The *Golapada* forms the fourth and the last part of *Aryabhatiyam*. It contains 50 stanzas. Important geometrical (and trigonometric) aspects of the celestial sphere are discussed in *Golapada*. The important features of the ecliptic, the celestial equator, the node, the shape of the earth, the cause of day and night, rising of the zodiacal signs on the eastern horizon, etc. find a place in this last part of the text. In fact, much of the contents of the *Golapada* of the *Aryabhatiyam* are generally discussed under a chapter called *triprasna* (three problems of time, place and direction) in the later *siddhantic* texts. Another very important topic included in this chapter is on the lunar and solar eclipses.

The system of astronomy expounded in the *Aryabhatiyam* is generally referred to as the *audayika* system since the *Kali* beginning is reckoned from the mean sunrise (*udaya*) at *Lanka* a place on the earth's equator. However, we learn from Varahamihira and Brahmagupta that Aryabhata I propounded another system of astronomy called *ardharatrika* in which the day is reckoned from the mean midnight (*ardharatri*) at Lanka. The important parameters are different in the two systems. However, Aryabhata's text of the *ardharatrika* system is not available now. Its parameters can be recovered from Brahmagupta's *Khandakhadyaka* (Sengupta 1934; Chatterjee 1970) and some later works. The beginning of *Kaliyuga* is (i) the mean sunrise (at Ujjayini) on February 18, 3102 BC (Julian) in the *audayika* system and (ii) the mean midnight between 17th and 18th, February, 3102 BC (Julian) in the *ardhratrika* system. The following are some of the innovative contributions of Aryabhata I:

1. A unique method of representing huge numbers using the alphabets for the purposes of metrics (*chandas*) and easy memorization. The method followed by Aryabhata is different from the now popular methods of *Katapayadi* (letter numerals) and *Bhutasankhya* (word numerals) which also serve the same purpose. However, Aryabhata's method was not followed by later astronomers perhaps due to the inconvenience of pronunciation, lack of meanings of the words formed and less choice than in the *Katapayadi*.
2. Sine tables: The importance of trigonometric functions like sine (*jya*) and cosine (*kotijya*) in Indian astronomy can hardly be exaggerated.

Aryabhata I gives the rule for the formation of the sine table just in one stanza! Accordingly, the sine values for the angles from 0° to 90° at intervals of $3°45'$ can be obtained. The values thus obtained compare well with the modern values. It is important to note that for an angle θ, the "Indian–sine" (*Jya*) of the angle θ is related to the modern sine by the relation

$$Jya\ (\theta) = Rsin\theta \tag{1.2}$$

where R is a predefined constant value of the radius of a circle. For example, Aryabhata, as also *Suryasiddhanta*, take the value $R = 3438'$ so that

$$Jya\ (\theta) = 3438' sin\theta \tag{1.3}$$

It is significant that $3438'$ is the value of a **radian** correct to the nearest integer. Brahmagupta takes $R = 150'$ and Bhaskara II has $R = 120'$ respectively in their *karana* texts. Any value can be chosen for R.

1.5 Earth's Shape and Rotation

Now, it is well known that the earth is spherical (or spheroidal) in shape and that it rotates about its own axis once a day causing day and night. Aryabhata I clearly mentions that:

1. The earth is spherical – "circular in all directions" (see *Golapada*, 6): *bhugolah sarvato vrtttah.*
2. Halves of the globes of the earth and the planets are dark due to their own shadows; the other halves facing the sun are bright. It is truly creditable that Aryabhata recognised that the earth and the other planets are not self-luminous but receive and reflect light from the sun (*Golapada,* 5):
   bhugrahabhanam golardhani svacchayaya vivarnani |
   ardhani yathasaram suryabhimukhani dipyante ||
3. Again, Aryabhata was the first to state that the rising and setting of the sun, the moon and other luminaries are due to the relative motion caused by the rotation of the earth about its own axis once a day. He says, "Just as a man in a boat moving forward sees the stationary objects (on either side of the river) as moving backward, just so are the stationary stars seen by the people at Lanka (i.e. on the equator) as moving **exactly** towards the west" (*Golapada*, 9).

The period of one sidereal rotation (i.e. with reference to the fixed stars in the sky) of the earth, as given by Aryabhata works out to be $23^h56^m4.1^s$. The corresponding mean modern value is $23^h56^m4.091^s$. The accuracy of Aryabhata is truly remarkable.

Aryabhata I (476 AD) is regarded as the major expounder of systematic and scientific astronomy in India. The unparalleled popularity of Aryabhata I and his system of astronomy is demonstrated by the fact that the remarkable development of astronomy in Kerala during 14th–19th centuries is based exclusively on the Aryabhatan system.

1.6 Post-Aryabhatan Astronomy

The cryptic and aphoristic style of Aryabhata would have made it extremely difficult to understand his text but for the detailed exposition of the system by Bhaskara I (c. 600 AD). In his commentary on the *Aryabhatiyam* as also in the works *Maha* (Shukla 1976) and *Laghu* (Shukla 1960) – *Bhaskariyams*, Bhaskara I (to be distinguished from his more popular namesake of the 12th century) has very ably expounded Aryabhata's astronomy with examples and copious references.

As mentioned earlier, Varahamihira (505 AD) brought together five systems of astronomy, extant during his period, in his remarkable work, *Pancasiddhantika* (Sarma 1993). He mentions that among the five systems, the *Suryasiddhanta* is the best. Varaha's *Suryasiddhanta* is now generally referred to as *Saurasiddhanta* to distinguish it from the later popular version of the *Suryasiddhanta* (Shukla 1957; Burgess 1989). It is believed that the *Suryasiddhanta* in its modern version was composed around 1000 AD. The parameters in the two texts are totally different.

While the *siddhantas* proper are large texts consisting of broad theories and a large number of topics, generally these texts are not handy for practical computations for day-to-day use. Further, very large numbers will have to be dealt with which are very inconvenient and lead to errors. Therefore, besides these *siddhantas*, two other genres of texts on astronomy have been in vogue. These are called *tantras* and *karanas*.

Conventionally, *siddhantas* chose the beginning of the *Mahayuga* (43,32,000) years or of the *Kalpa* (432×10^7 years) as the epoch. After the *Suryasiddhanta* two famous *siddhantas* are *Brahmasphutasiddhanta* (Sharma 1966) of Brahmagupta (628 A.D.) and *Siddhantasiromani* (Dvivedi 1929) of Bhaskara II (1150 AD). A large number of commentaries and even super-commentaries are written particularly on the *Suryasiddhanta*.

The *tantra* texts have comparatively fewer topics and explanations. These works choose the more convenient epoch viz. the beginning of the *Kaliyuga* (the midnight of 17/18 February 3102 BC or the sunrise of February 18). For example the *Aryabhatiyam* (Shukla and Sharma 1976; Shukla 1976; Sastri 1977) and Nilakantha Somayaji's *Tantrasangraha* (Sarma 1977) (c. 1500 AD) are *tantra* texts.

However, for practical computations and making *pancangas* the most useful handbooks are the *karana* texts. In each of these, practical algorithms are provided taking a convenient contemporary date as the epoch. The advantage of a recent epoch is that one now deals with smaller numbers for the *ahargana* (the number of civil days elapsed since the epoch). Further, since corrected positions of planets for a recent date will have been given with necessary *bijasamskaras* (corrections), the computations based on these *karana* handbooks are more accurate. The well-known *karana* texts are Brahmagupta's *Khandkhadyaka* (7th century) (Sengupta 1934; Chatterjee 1970), Bhaskara II's *Karanakutuhalam* (12th century) (Rao and Uma 2007–08) and Ganesa Daivajna's *Grahalaghavam* (epoch: 1520) (Rao and Uma 2006). A large number of such handbooks and tables (*saranis*) were composed during different periods, even as late as in the 19th century These include the remarkable works of Kerala astronomers: Madhava, Paramesvara, Nilakantha Somayaji, Jyesthadeva and Acyuta Pisarati.

1.7 Contents of *Siddhantas*

Various topics of interest in Indian astronomy are discussed in different chapters. A chapter is called *adhyaya* or *adhikara*. The following are the topics discussed in the different *adhikaras* in a typical *siddhantic* text.

1.7.1 Madhyamadhikara (Mean Positions)

The word *madhyama* means the average or 'mean' positions of the heavenly bodies. These include the sun, the moon and the so-called *taragrahas* viz. Mercury, Venus, Mars, Jupiter and Saturn. In order to calculate the mean angular velocities, the numbers of revolutions completed in a *Mahayuga* (of 43,20,000 years) or a *Kalpa* (432×10^7 years) by the bodies as also by the special points viz. the apogee (called *mandocca*) of the moon and the moon's ascending node (popularly called *Rahu*) are given (Rao 2004).

The procedure to calculate the *ahargana* (the number of civil days from the epoch) for a given date is also explained in this chapter. The total number of civil days in a *Mahayuga* (MY) is also specified. Then, the motion of a planet from the epoch to the given date is given by:

$$\text{Motion} = \frac{\text{No. of revolutions in a MY} \times \textit{Ahargana} \times 360^\circ}{\text{No. of civil days in a MY}} \text{ in degrees} \tag{1.4}$$

When the nearest integral multiple of 360° (i.e. completed number of revolutions) is dropped from the above value, we get the mean position of the planet in degrees, etc. for the given date. The revolutions of the sun, the moon, the moon's apogee and node in a *Kalpa* (432×10^7 years), according to different texts are given in Table 1.3. The last column in the table gives the civil days in a *Mahayuga* (432×10^4 years). In fact this number of civil days defines the length of a sidereal solar year:

$$\textit{Sidereal (nirayana) solar year} = \frac{\text{Number of civil days}}{43{,}20{,}000} \tag{1.5}$$

For example, according to the *Suryasiddhanta*,

$$\textit{Nirayana} \text{ solar year} = \frac{1{,}57{,}79{,}17{,}828}{43{,}20{,}000} = 365.258756481 \text{ days} \tag{1.6}$$

From Table 1.3 we notice that while the revolutions of the sun in a *Kalpa* are the same according to the different texts, those of the moon, its apogee and node *(Rahu)* are different. Further, the number of civil days in a *Mahayuga* (or multiplied by 1000 for a *Kalpa*) are slightly different according to Aryabhata I and the *Suryasiddhanta*, for example. These differences have resulted from the corrections (*bijas*) made periodically, and give rise to slightly different mean daily rates of motions.

The mean daily motions are given in Table 1.5 according to the *Siddhantasiromani* of Bhaskara II as compared to the modern values and those of the *Khandakhadyaka* of Brahmagupta and the *Suryasiddhanta*. The values are given in degrees, minutes, seconds of arc and two further subdivisions of a second viz. 1/60 and 1/3600 of second of arc.

The revolutions of the planets (*taragrahas*), according to the *Suryasiddhanta*, are listed in Table 1.4

Let λ_o be the mean longitude of a planet at the chosen epoch, d be the daily mean motion of the planet and A be the *Ahargana,* the number of days elapsed since the epoch up to the day under consideration. Then, the mean longitude of the planet is given by

Table 1.3 Revolutions of the sun, the moon, etc. in *Kalpa* (1 *Kalpa* = 432 × 10^7 years = 1000 *Mahayugas*)

Bodies and points	*Ravi*	Candra	Candra's Apogee (*Mandocca*)	Candra's (Asc.) Node (*Rhu*)	Civil days in a *Mahyuga* of 432 × 10^4 yrs.
Aryabhata I	4,32,00,00,000	57,75,33,36,000	48,82,19,000	23,21,16,000	1,57,79,17,500
Brahmagupta *Khandakhadyaka*	432,00,00,000	5775,33,36,999	48,82,19,000	23,22,26,000	1,57,79,17,800
Suryasiddhanta	4,32,00,00,000	57,75,33,36,000	48,82,03,000	23,22,38,000	1,57,79,17,828
Aryabhata II (*Mahasiddhanta*)	4,32,00,00,000	57,75,33,34,000	48,82,08,674	23,23,13,354	1,57,79,17,542
Bhaskara II (*Siddhantasiromani*)	4,32,00,00,000	57,75,33,00,000	48,82,05,858	23,23,11,168	1,57,79,16,450

$$\lambda = \lambda_o + A \times d \tag{1.7}$$

As in the case of the sun and the moon, we choose the beginning of *Kaliyuga*, i.e. the mean midnight between 17 and 18 February 3102 BC as the epoch. The mean daily motion (in revolution) is given by:

d = No. of revolutions in a *Kalpa*/No. of civil days in a *Kalpa*

where 1 *Kalpa* = 432 × 10^7 years. The details are given in Table 1.3. In the case of *Budha* and *Sukra* the positions of their *Sighrocca* are considered. In fact of these are the mean Mercury and Venus according to modern theory. The mean daily motions of the heavenly bodies and special points, according to different texts are shown in Table 1.5. along with those according to modern astronomy for comparison.

Table 1.4 Revolutions of planets in a *Mahayuga* (*Suryasiddhanta*) (Number of civil days in a *Mahayuga 1,57,79,17,828)*

Planet	No. of revolutions	Mean daily motions (d)
Kuja	22,96,832	0°.5240193
Budha-sighrocca	1,79,37,060	4°.0923181
Guru	3,64,220	0°.0830963
Sukra Sighrocca	70,22,376	1°.6021464
Sani	1,46,568	0°.0334393

1.7.2 Spastadhikara (True Positions)

In this chapter the procedure to obtain the "true" position of a planet, from the mean position, is discussed. The word *spasta* means correct or true. For obtaining the true positions from the mean, two corrections are prescribed:

1. *Manda*, applicable to the sun, the moon and the five planets and
2. *Sighra,* applicable only to the five planets (*taragrahas*) viz. *Budha, Sukra, Kuja, Guru* and *Sani.*

The *manda* correction takes into account the fact that the planets' orbits are not circular. This correction corresponds to what is called "the equation of the centre" in modern astronomy. The *sighra* correction corresponds to conversion of the heliocentric positions of planets to the geocentric in the case of the five planets. It is significant that ancient Indian astronomers applied this equation, correctly, without realizing that the motion was heliocentric.

Table 1.5 Mean daily motions of the sun, the moon, etc.

Bodies and points	*Khandakhadyaka*			*Suryasiddhanta*					*Siddhantsiromani*					Modern astronomy		
	o	*i*	*ii*	o	*i*	*ii*	*iii*	*iv*	o	*i*	*ii*	*iii*	*iv*	o	*i*	*ii*
Ravi	0	59	08	0	59	08	10	09.7	0	59	08	10	21	0	59	08.2
Candra	13	10	31	13	10	34	52	02	13	10	34	53	00	13	10	34.9
Kuja	0	31	26	0	31	26	28	10	0	31	26	28	07	0	31	26.5
Budha's Sighrocca	4	05	32	4	05	32	20	42	4	05	32	18	28	4	05	32.4
Guru	0	04	59	0	04	59	08	48	0	04	59	09	09	0	04	59.1
Sukra's Sighrocca	1	36	07	1	36	07	43	37	1	36	07	44	35	1	36	07.7
Sani	0	02	00	0	02	00	22	53	0	02	00	25	51	0	02	00.5
Candra's Mandocca	0	06	40	0	06	40	58	42	0	06	40	53	56	0	06	40.92
Candra's Pata (Rahu)	–0	03	10	–0	03	10	44	43	–0	03	10	48	20	–0	03	10.77

1.7.3 Triprasnadhikara

This chapter deals with the "three questions" of direction (*dik*), place (*desa*) and time (*kala*). Procedures for finding the latitude of a place, the times of sunrise and sunset, variations of the points of sunrise and sunset along the eastern and western horizons, gnomon problems *kranti* (declination) and calculation of *lagna* (ascendant) are discussed. Traditional Indian astronomers have shown their remarkable mastery over spherical trigonometry.

1.7.4 Candra and Surya Grahanadhikara

In these two chapters the computations of the lunar and the solar eclipses are discussed. The instants of the beginnings, the middle and the endings, effects of parallax, regions of visibility, possibility of the occurrence, totality, etc. of the eclipses are considered. Their computational procedures are elaborated. Improvements in procedures and updating of parameters are effected by great savants like Samanta Chandrasekhara Simha (Upadhyaya 1996, 1998), Venkatesha Ketkar (Ketkar 2006; Ketkar 2008) and T.S. Kuppana Sastri. The classical procedures and the improved procedures are discussed by S. Balachandra Rao and Padmaja Venugopal (Rao and Venugopal 2008).

In fact, for Indian astronomers, the true testing ground for the veracity of their theory, parameters and procedures very much depended on the successful and accurate predictions of eclipses. Of course, as and when minor deviations between computations and observations were noticed, necessary changes and corrections (*bijasamskara*) were suggested to achieve concurrence of observations and computations (*drgganitaikhya*).

Besides these four important topics, the *siddhantic* texts contain many other topics, which vary from text-to-text, like the first visibility of planets, moon's cusps, mathematical topics like *Kuttaka* (solution of indeterminate equations), spherical trigonometry and the rationales of the formulae used, etc. Some such important chapters in the *siddhantic* texts are:

(i) *Udayastadhikara*: Rising and setting of the sun, the moon, the planets and stars, daily as well as heliacal.

(ii) *Grahayutyadhikara*: Conjunctions of two heavenly bodies. These include transits and occultations also.

(iii) *Srngonnati adhikara*: The altitudes of the horns of the moon.

Different texts discuss some more topics and sometimes the distribution of topics into chapters is also different.

1.8 Continuity in Astronomical Tradition

A characteristic feature of Indian astronomy is the unbroken continuity in the tradition, starting from Vedic period up to the colonial period. Starting from simple observations and a simple calendar, relevant to the contemporary needs during the Vedic times, there has been a gradual progress in the extent of astronomical topics considered, mathematical techniques developed, and refinement and sophistication in the computational algorithms, always aimed at greater accuracy during the *siddhantic* period of evolution spread over nearly fifteen centuries. However, the progress from *Vedanga Jyotisa* to Aryabhata – during about the fifteen century – is not clear!

The existing popular *siddhanta* texts, like *Suryasiddhanta*, are made clearer with elucidations and illustrations by a large number of commentaries, super-commentaries, etc. For example, the *Aryabhatiyam* carries highly learned and exhaustive commentaries by Bhaskara I (Shukla 1976) Paramesvara (Kern 1990) and Nilakantha Somayji (Sastri 1977) among others. Prthudakasvamin's commentary (Sengupta 1934) on the *Khandakhadyaka* of Brahmagupta, besides those by Bhattopala (Chatterjee 1970) and Amaraja, is extremely useful. Bhaskara II has written his own commentary, *Vasana bhasya* on his magnum opus, *Siddhantasiromani* (Dvivedi 1929). In fact, often the commentaries improve upon the parameters and computational techniques of the original texts to yield better results.

While Manjula (Shukla 1990) (or Munjala 932 AD) and Sripati (Misra 1932) (c. 1000 AD) introduced additional corrections for the moon, Nilakantha Somayaji (c. 1500 AD) revised the model of planetary motion in his *Tantrasangraha* (Sarma 1977) for obtaining better positions of the inferior planets, *Budha* and *Sukra*.

Inspired by the ideas of Paramesvara (c. 1400 AD), Nilakantha (c. 1500 AD) developed a model in which all planets move round the sun in eccentric orbits and the sun moves round the earth. Today we know that actually it is the earth that goes round the sun. However, for an observer on the earth, due to relative motion, the sun *appears* to move round the earth. It is a significant achievement in Indian astronomy, before Copernicus came into picture. Nilakantha's revised model was successfully adopted by all later astronomers of Kerala, like Jyesthadeva, Acyuta Pisarati and Citrabhanu. Nilakantha's breakthrough in the planetary model is brought out by M. S. Sriram et al. 2002

It is also noteworthy that the knowledge of astronomy was never restricted to any particular region, but spread throughout India. After Bhaskara II (b. 1114) there was a great surge of mathematicians and astronomers of Maharashtrian origin like Kesava Daivajna, his son Ganesa Daivajna, Divakara, Visnu, Mallari, Visvanatha. While Candrasekhara Samanta of Orissa made quite a few important innovations, like an additional correction to the moon, independently, Kerala became the pocket of tremendous development during 14th–19th centuries. Of course, congenial social milieu and patronage must have played an important role in the development of astronomy more during certain periods and in certain regions and less at other times and regions.

In recent centuries, extensive study and research have gone into Indian classical astronomy. The great stalwarts who have contributed for such an in-depth critical study are P.C. Sengupta, Bina Chatterjee, Venkatesh Ketkar, T.A. Sarswathi Amma, T.S. Kuppanna Sastry, K.V. Sarma, K.D. Abhyankar and a few others.

1.9 Observational Validity of Indian Astronomical Parameters

Roger Billard in his publication of 1971 adopted the novel procedure of mathematical statistics, based on the method of least squares, analysed a large number of astronomical *siddhantas*, and graphically represented the deviations, against time, of longitudes, calculated according to the texts based on modern procedures and tables. From these deviation curves he determined accurately and independently the dates of compositions of a number of texts. Also, it was observed that the values of the planetary positions obtained by theoretical calculations concurred with observed values. The constants used for the calculation of planetary positions were upgraded to obtain the values comparable to the modern ones.

Billard recognised three periods in the development of Indian astronomy. The first period dated from the time of the *Brahmanas* (about 10th to 8th century BC) and extended up to 3rd century BC. Its characteristic feature was calendrical astronomy and the representative text was the *Vedanga Jyotisa*. The second period extended from 3rd century BC to the 1st century AD and was marked by Babylonian astronomical elements, particularly the *tithi* as a unit of time, arithmetical computations based on heliacal risings and settings of planets, their synodic revolutions, etc. The third period, the period of scientific astronomy, extended from around AD 400 up to modern times. Its important characteristic is the employment of trigonometrical methods and epicylic models for the computations of planetary positions, and production of the *Aryabhatiyam* of Aryabhata I (476 AD) as the first text of the kind. Billard is mainly concerned with the scientific astronomy of the third period because the relevant texts and materials available paved way for the investigation by the statistical method adopted by him.

Before applying the method of computation of planetary position given in Sanskrit astronomical texts, Billard first tested its reliability with respect to Ptolemy's *Almagest* (*Mathematical Syntaxis*). The epoch of the *Almagest* was taken to be February 26, 746 BC. Between this date and February 2, 141 AD (Julian) there were records of 76 observations, mostly of lunar eclipses and a few of planets. Ptolemy has given formulae for apogees (aphelions), eccentricities, radii of epicycles and conversion of mean longitudes to true longitudes. The graphical representation showed that the longitude deviation curves converge to a point corresponding to 100 AD, except for Mercury and Venus. The error is about nearly 1° taking into account Ptolemy's observational error. The year of zero error is 126 BC, and that agrees with Hipparchus' period.

Billard has discussed Aryabhata's *ardharatrika* and *audayika* systems. In the former, the epoch is the midnight at the beginning of the *Kaliyuga* where the *yuga* comprises 432×10^4 years of 1577917800 civil days, while in the latter the *Kaliyuga* starts at the following sunrise and the number of civil days is taken as 1577917500 days. The numbers of days elapsed in 3,600 years of *Kaliyuga* era on Sunday, March 21, 499 AD are 1314931.5 in the *ardharatrika* system and 1314931.25 using the *audayika* system with a difference of exactly 6 hours. We note that

$$\frac{1314931.5}{1577917800} = \frac{1314931.25}{1577917500} = \frac{3600}{4320000} = \frac{1}{1200} \tag{1.8}$$

Billiard in his work, with the help of modern statistical-cum-computer methods has understood the true nature, the merit, the originality and the shortcomings of the Sanskrit astronomical system and has shown that it functioned as an active and effective institution for over fourteen centuries and maintained its vitality through periodical observations and emendations. Billard's approach is a contrast to the general efforts of the biased western scholars directed at digging evidence in favour of their pet theory of the imported character of the system, its lack of originality, its absence of an observational basis and so on.

On David Pingree's theory of Aryabhata's dependence on Greek planetary tables, Van der Waerden (1980) considers it practically impossible for the following reason. The most accurate Greek tables were the *Handy Tables* of Ptolemy. They were used in Alexandria and Byzantium in the time between Ptolemy and Aryabhata. As far as longitudes are concerned, these tables are based on the *Almagest*. Without any doubt pre-Ptolemaic Greek tables would give rise to still larger deviations. The errors of astronomical tables normally tend to increase in the course of time. By sheer accident, it may happen that for one or two planets the deviations become nearly zero for a particular year. But Waerden has shown graphically that deviation of nine planets simultaneously become extremely small in 510 AD. The only possible explanation is that mean longitudes in the *Aryabhatan* system were determined by accurate observations around 510 AD.

Supposing Aryabhata had tables from which mean and true positions could be computed for his own time and he had really computed them to find discrepancies between the observed and calculated positions, how would he proceed to make his system work Vander Waerden says that, "he might have made appropriate changes in the elements of the table and calculated the positions anew". After one or two trials, he would probably get an agreement between the observed position and calculated value. There was no need for him to calculate mean longitudes from observed positions. That Aryabhata really proceeded in this manner is strongly indicated in Billard's analysis of Aryabhata's two systems (midnight and sunrise systems).

In the *ardharatrika* system all mean longitudes are comparable for 510 AD except for Mercury and Jupiter. Supposing Aryabhata wanted to improve upon his Jupiter, it would have been easier for him to observe the planet for sometime, say one or two synodic periods of 13 months each, and see what changes are to be made in the mean longitudes and other elements in order to get better values which are comparable between the observed and the computed ones.

According to Vander Waerden, Sanskrit astronomical works do not suggest backward computation of conjunction, but Persian sources as preserved in the writings of later Arab astronomers do. In his *Book of Conjunctions,* Abu Ma'shar used Persian sources for mean motions of Saturn and Jupiter and showed that 18,138 mean conjunctions took place in 360,000 years, that is, one conjunction is nearly 20 years, and the mean motion from one conjunction to the next was $242°25'17''$.

Observational basis is also strongly indicated in Billard's analysis of Brahmagupta's *Brahmasphutasiddhanta.* Although the text was written in 628 AD the planetary positions recorded in the system appear to have been made near the middle of the 6th century and those of the sun and the moon near the end of the century. Vander Waerden carried out his calculations to confirm these conclusions. Aryabhata's mean longitudes which were accurate for 510 AD. would deteriorate progressively and show up large errors in the case of Mars, Saturn, Jupiter and Venus after nearly a century. Brahmagupta, or his predecessor gave corrections to the planetary positions which after incorporating gave the correct values. This proves that the corrections made for various planets were based on observations and the corrections were given same time after Aryabhata's period.

In the traditional *siddhantic* texts, the computations of true sun, moon and three of the five planets come out quite well. But, Mercury and Venus somehow eluded accurate computations. The famous Kerala astronomer Nilakantha Somayaji asserted that these two inferior planets must be treated on par with the remaining planets in so far as determining their *Sighraphala* (equation of conjunction). In the earlier texts the practice was to consider two fictitious points as *Sighroccas* (apex of conjunction) in the case of Mercury and Venus while the mean sun was taken as the *Sighrocca* of the superior planets. In fact, Nilakantha's treatment of all the planets the same way, in this context, by considering the sun as the apex of conjunction is a very significant breakthrough in the Indian astronomical development. Nilakantha's theory enunciated in his celebrated work, *Tantra Sangraha* (around 1500 AD) hints at a heliocentric model of planetary motion. This important aspect has been highlighted by K. Ramasubramaniam, M.D. Srinivas and M.S. Sriram (1994) in their paper (Ramasubramanian et al. 1994).

1.10 Indian Astronomers and Eclipses

In ancient and medieval astronomical texts (*siddhantas*, *tantras* and *karanas*) great importance is given to the phenomenon and computation of eclipses (*grahana, uparaga*). The Indian astronomers used to put to test their theories and computations in respect of positions of the heavenly bodies – especially the sun and the moon – on the occasions of the eclipses. As and when disagreements occurred between the observed and the computed positions, the great savants of Indian astronomy revised their parameters and when necessary even the computational procedures. Improving the computations of eclipses-based on sustained observations over long periods of time – was an important target of *Siddhantic* astronomers.

The scholarly Kerala astronomer, Nilakantha Somayaji (1444–1545) paying glowing tributes to his *parama guru* (grand-teacher), Paramesvara (1362–1455), remarks "*Paramesvara . . . having observed and carefully examined eclipses and conjunctions for 55 years composed his Samadrgganitam* [*Pancapancasat varsa-kalam niriksya grahana grahayogadisu|pariksya samadrgganitam karanam cakara*]."

After Bhaskara II (b. 1114 AD) there was an apparent decline in mathematical and astronomical output in India; however, after some time there was a tremendous development in Kerala due to stalwarts like Madhava, Paramesvara and Nilakantha Somayaji from about 14th–17th centuries of the Christian era.

During that period were quite a few astronomers of great achievement in other part of India also. In fact, among the astronomical works in common us, especially in *pancanga*-making, Ganesa's *Grahalaghavam* (Rao and Uma 2006) is the most popular one. Even to this day in most of northern India, Maharashtra, Gujarat and north Karnataka, the *Grahalaghavam* (GL) is in vogue.

While the geometrical configurations and mathematical procedure for computations of eclipses presented by the traditional Indian astronomers are quite sound, the relevant parameters, over a period of centuries, need to be upgraded periodically. With this view, have the present author and his associates developed such a procedure and presented the same through traditional as well as modern examples (Rao and Venugopal 2008). This procedure is inspired by the works of the reputed savant, the later Kuppanna Sastri of Madras (*Chennai*) (Sastri 1989).

1.10.1 Real Cause of Eclipses

The real scientific causes of the lunar and solar eclipses were well known to the ancient Indian astronomers even in the pre-Aryabhatan period. The procedure explained in the then extant *panca siddhantas* (Sarma 1993) (five systems of astronomy) bear this out. Aryabhata I (b. 476 AD) explains the causes of the two types of eclipses and explain in his characteristic style of brevity:

chadayati sasi suryam sasinam mahati ca bhurrhaya

"The moon covers the sun and the great shadow of the earth (eclipses) the moon."

-*Aryabhatiam* (Shukla and Sharma 1976) 4, 37

Varahamihira (c. 505 AD) explains at length in his *Brhat Samhita* the real causes of the eclipses and debunks the irrational myths generally entertained by the ignorant masses. He declares:

bhucchayam svagrahane bhaskaram arkagrahe pravisati induh

"At a lunar eclipse the moon enters the shadow of the earth and at a solar eclipses the moon enters the sun's disc".

- *Br. Sam,* (Bhat and Banarasidass 1981) 5, 8

Varahamihira gives all credit to the ancient preceptors for the knowledge of the causes of eclipses in saying:

evam uparaga karanam uktamidam
divyadrgbhih acaryaih
rahurakaranam asmin niyuktah
sastra sadbhavah

"In this manner, the ancient seers endowed with divine insight have explained the cause of eclipse (*uparaga*). Hence the scientific fact is that (the demon) *Rahu* is not at all the cause of eclipse."

- *Br. Sam,* 5, 13

Further, demolishing some of these superstitious beliefs regarding eclipses, Varaha remarks:

na kathancidapi nimittaih grahanam vijnayate nimittani |
anyasminnapi kale bhavanti atha utpata rupani ||

"An eclipse can by no means be ascertained through omens and other indications. Because the portents such as the fall of meteors and the earthquakes occur at the other times also."

- *Br. Sam,* 5, 16

He also says:

pancagraha samyoganna khila grahanasya sambhavo bhavati
tailam ca jale astamyam na vicintyamidam vipascidbhih

"Scholars should not believe in the following myth of the ignorant to the effect that an eclipse cannot take place except when there is a combination of five planets in the same zodiacal sign, and that a week before the eclipse, i.e. on the preceding 8th lunar day, its characteristics can be inferred from the behaviour on appearance of a drop of oil poured on the surface of water."

- *Br. Sam,* 5, 17

1.10.2 Moon's Latitude

Generally, in traditional Indian astronomical texts, *mean Rahu* and its *mean* daily motion are used. But actually we will have to consider the *true Rahu* and its *true* daily motion for determining the *true* latitude (*viksepa*) of the moon. However, the error caused is not significant since at an eclipse, the moon is close to one of the nodes (Rao et al. 2003).

According to *Khandakhadyaka* (KK), we have Moon's latitude (*viksepa*) given by

$$\begin{aligned} Viksepa &= [Jya\ (moon\text{-}Rahu)] \times \tfrac{9}{5}\ minutes\ of\ arc \\ &= [150' sin(M-R) \times \frac{9}{5} = 270' sin(M-R) \end{aligned} \tag{1.9}$$

1.10.3 Diameters of Sun, Moon and Earth's Shadow

Brahmagupta next describes the determination of the angular diameters (*bimba*) of the sun, the moon and the earth's shadow (*chaya*):

> *bhavadasagunite ravisasigati nakhaih svarajinairhrte mane* |
> *sastya bhaktam tatvastagunitayor antaram tamasah* ||
>
> The (true daily) motions of the sun and the moon multiplied (respectively) by 11 and 10 and divided by 20 and 247 are their measures (diameters in arc-minutes). Multiplication (of the true daily motions of the sun and the moon) by 25 and 8 (respectively) and division of their difference by 60 is that (diameter) of the shadow.
>
> - *KK* (Sengupta 1934; Chatterjee 1970) IV, 2

Explanation: *The bimba* of the sun, the moon and the earth's shadow are given by the following expressions
(i) Sun's diameter = $\frac{11}{20} \times SDM$ minutes
(ii) Moon's diameter = $\frac{10}{247} \times MDM$ minutes
(iii) Shadow's diameter = $(28 \times MDM - 25 \times SDM)/60$ minutes
Rationale: It is assumed that the *true diameter* of the sun or the moon's *proportional* to the *true daily motion* of the concerned body. Thus,

True diameter = $\frac{\text{Mean diameter} \times \text{True daily motion}}{\text{Mean daily motion}}$

We have

(i) The mean diameter of the sun = $32'31''$
The mean daily motion of the sun = $59'8''$
True diameter of the sun=$\frac{32'31''}{59'8''} \times SDM \approx \frac{11}{20} \times SDM$
(ii) The mean diameter of the moon = $32'$
The mean daily motion of the moon = $790'31''$
True diameter of the moon = $\frac{32'}{790'31''} \times MDM \approx \frac{10}{247} \times MDM$
(iii) **Angular diameter of the shadow:** Let S and E be the centres of the sun and the earth and M be the position of the moon. In Figure 1.1 the sun's horizontal parallax = $E\hat{B}M$ and the moon's horizontal parallax = $E\hat{M}B$.
Now, in triangle BEM, we have

$$E\hat{B}M + E\hat{M}B = M\hat{E}B' = M\hat{E}V + V\hat{E}B'$$

But $V\hat{E}B' = S\hat{E}B$, the sun's angular semi-diameter (*Ravi bimbardha*).

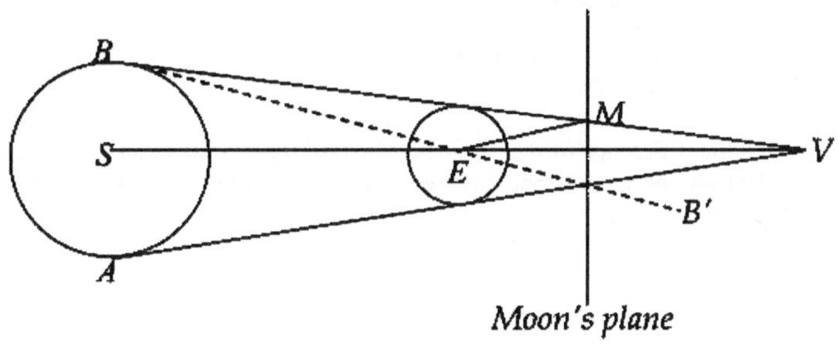

Fig. 1.1 Angular diameter of the shadow

Therefore, the angular diameter of the shadow
$= 2M\hat{E}V = 2[E\hat{B}M + E\hat{M}B - S\hat{E}B]$ at the plane of the moon's orbit.
For example the angular diameter of the shadow
= 2[sun's horizontal parallax + moon's horizontal parallax− sun's angular diameter]
$= 2\left[\frac{SDM}{15} + \frac{MDM}{15} - \frac{11}{40} SDM\right.$
$= (1/60)[8\,SDM + 8\,MDM - 33\,SDM]$
$= (1/60)[8\,MDM - 25\,SDM]$
where MDM = Moon's true daily motion and SDM = sun's true daily motion.

Note: Indian astronomers have assumed that the horizontal parallax of the body is equal to the true daily motion of the body divided by 15. While this is somewhat true in the case of the moon, it completely fails in the case of the sun and planets.

1.10.4 Grasa (or Channam) - Obscured Portion

Now, Brahmagupta describes the "obscured portion" (*grasa*) as follows:

viksepam samsodhya pramanayogardhatastamaschannam |
sarvagrahanam grahyadadhike khandgrahanamune ||

Subtracting the (moon's) latitude from half the sum of the diameters (of the obscured and the obscuring bodies), the remainder is the *channam* (or *grasa*), the obscured portion. If the *channam* (obscured portion) is *greater than* the (diameter of) obscured body the eclipse is *total*; if less then *partial*.

- *KK*, IV, 3

Explanation:
Obscured portion
$= \frac{1}{2}$[Diameter of the eclipsed body + Diameter of the eclipser]$-$Latitude
(i) In lunar eclipse,
Obscured portion (*channam*) $= \frac{1}{2}[MDIA + SHDIA] - \beta$
(ii) In solar eclipse,
Obscured portion (*channam*) $= \frac{1}{2}[SDIA + MDIA] - \beta$
where *MDIA*, *SHDIA* and *SDIA* are respectively the diameters of the moon, the earth's shadow and the sun; β is the latitude of the moon at the instant of the opposition and the conjunction of the sun and the moon at lunar and solar eclipses, respectively.
Note: $\frac{1}{2}$[Diameter of the eclipsed + Diameter of the eclipser] is called *Manaikya khandardha*.
Note: (i) If $\frac{1}{2}[MDIA + SHDIA] < \beta$, i.e. *channam* is *negative*, there will be *no eclipse.*
(ii) If $channam < MDIA$, the eclipse is *partial*; and
(iii) If $channam > MDIA$, the eclipse is *total.*

1.10.5 Half-durations of Eclipse and Totality

Next, Brahmagupta proceeds to explain how to compute the half-durations of the eclipse and of the totality:

chadyardhena cchadakadalasya yuktonkasya vargabhyam |
viksepakrtim prohya pade tithivat sthitivimardardhe ||

From the squares of the sum and the difference of the semi-diameters of the obscured and the obscuring bodies subtract the square of the moon's latitude; from the square-roots (of the two results), in the same way as of *tithis*, the half-durations of the eclipse and of the totality (are obtained).

- *KK*, IV (Sastry 1993)

Explanation: In a lunar eclipse, we have
(i) Half-duration of the eclipse (*sthiti-ardha*)

$$= \left[\sqrt{\left[\frac{1}{2}(MDIA + SHDIA\right]^2 - \beta^2} \div (MDM - SDM)\right] \times 60\,gh.$$

(ii) Half-duration of totality (*Marda-ardha*)

$$= \left[\sqrt{\left[\frac{1}{2}(SHDIA - MDIA)\right]^2 - \beta^2} \div (MDM - SDM)\right] \times 60\,gh.$$

In a solar eclipse, *MDIA* and *SHDIA* have to be replaced respectively by *SHDIA* and *MDIA*.

1.10.6 Successive Approximation for Circumstances

Brahmagupta, being a mathematician and observational astronomer *par excellence* aims at accuracy by the process of 'iteration'.

bhuktih sastihrta sthitivimardadalanadikagunarkendvoh |
adavrnamante dhanamasakrt tenanyatha pate ||

The true daily motion of the sun and the moon divided by 60 and multiplied by the half-durations of the eclipse and of the totality must be **repeatedly applied**, negatively for the beginning and positively for the end (of the eclipse and of the totality); for the node this is (applied) in the reverse order.

-*KK* (Shukla and Sharma 1976) IV

Explanation: The half-durations of the eclipse obtained above (from sloka 4) are only a *first approximation*. The beginning and the end of the eclipse (and of totality obtained by subtracting from and adding to the instant of opposition, respectively, are also first approximation. This is so since, to start with, we do not know the precise instants and hence the latitude of the moon at those instants; we will have used the moon's latitude at the instant of the opposition as a first approximation.

At the thus obtained beginning and ending moments, the true positions of the sun, the moon, *Rahu* and hence the moon's latitude are determined. From these again the half durations of the eclipse (and of totality) are found out as a second approximation. This process is repeated successively till the values in each circumstance *converge*. Thus, by *successive approximation* the actual instants of the beginning and the end of the eclipse (and of totality) are determined. This mathematical procedure of iteration is called "*asakrt*".
Note: In modern astronomy also, the actual instants of the four "contacts" are determined by *successive approximation*.

1.11 Conjunction of Stars and Planets

In Indian Astronomy, the ***siddhantic*** texts have discussed in detail the phenomenon of conjunctions of the sun, the moon and the planets, between any two of them as also with some important stars like *Rohini* (Aldebaran), *Tisya* (i.e. Pusya, δ Cancri), *Citra* (Spica), etc.

In this section we discuss the actual working of the transit of Venus and Mercury by the method of (improved) Indian classical astronomy. The results are comparable to those of modern astronomical procedures (Rao and Venugopal 2009; Rao et al. 2008).

1.11.1 Types of Conjunctions

Among the planets (*taragrahas*), conjunction (*yuti*) takes place between any two of them and also with the moon or the sun. This phenomenon takes place when the longitudes of the two concerned bodies are equal and their latitudes are within permissible limits of difference even as in the case of the eclipses of the sun and the moon. The *Suryasiddhanta* describes the same as follows:

tara grahanamanyo'nyam syatam yuddhasamagamau |
samagamah sasankena suryenastamanam sayha ||

If the conjunction is (1) between two planets it is called *yuddha* (encounter); (2) between a planet and the moon, then it is called *samagama*. In modern parlance it is called **lunar occultation**; (3) between a planet and the sun, it is called *astamana* or *astangata* (heliacal setting).

- *Sury Siddh,* Ch. 7, *Sl.* 1

In particular if Mercury or Venus is in conjunction with the sun, such an alignment is called "**transit**". The transit of Venus on June 8, 2004 was truly spectacular.

1.11.2 Conjunction – Elapsed or to Occur

Sighre mandadhike'titah samyogo bhavitanyatha |
tayoh pragyayinorevam vakrinostu viparyayat ||

If the longitude of the faster moving body is greater than that of the slower one, the conjunction *(samyoga)* is passed. Otherwise the conjunction is yet to happen. This is the case when the two bodies are moving eastward (direct motion). However if both the bodies are retrograde (*vakri*) the reverse order is followed.

- *Surya Siddh*, Ch. 7 *Sl*, 2

1.11.3 Instant of Conjunction of Two Bodies

The famous Indian astronomer of the early 16th century, Ganesa Daivajna (epoch: March 19, 1520) describes the method of determining the instant of conjunction and also whether it is past or yet to occur in his *Graha-laghavam* (GL):

khrujugatigayostu vakrayorva vivarakala gatijantarena bhaktah |
gatijayutihrta yadaikavakri yutiragata pragataptavsaraih syat ||

This *sloka* explains how to obtain days (and *ghatis,* etc.) of *gata* or *gamya* instant of conjunction.

- *GL,* Ch. 13, Sl.3*l.* 3*[1]

1. If both the bodies are in direct motion or in retrograde motion then divide the difference between their positions by the difference between their daily motions, both the values expressed in the same unit of angle viz. *kala* (minute) or *amsa* (degree). The result will be in days, *ghatis,* etc. This gives the time in days, etc. of the *gata* (elapsed) or *gamya* (to be covered).
2. Of the two bodies if one is in direct motion and the other in retrograde motion then divide the difference between their positions by the *sum* of their daily motions.

Example: Suppose the longitude of *Kuja*, $P_1 = 10^R\,6°35'9''$ and the longitude of *Sani*, $P_2 = 10^R\,2°58'44''$ and both are in direct motion. The number of days for *yuti* is given by $\frac{P_1-P_2}{DP_1-DP_2}$ where the difference between the longitudes of *Kuja* and *Sani*

$P_1 - P_2 = 3°36'25'' = 216'25''$ and the difference of their daily motions

$$DP_1 - DP_2 = 42'50' - 3'3'' = 39'47''. \tag{1.10}$$

$$Days\ for\ yuti = \frac{216'25''}{39'47''} = 5^d 26^{gh} 23^{vig} \tag{1.11}$$

Since the longitude of the faster *Kuja* is geater than that of *Sani*, the *yuti* is over by $5^d 26^{gh} 23^{vig}$ before the given date.

1.11.4 Bheda and Abheda Conjunctions

Even as in the cases of solar and lunar eclipses, two bodies may have the same longitude but those may not form technically a "*bhedayuti*" in the sense of one body eclipsing the other or a "close" conjunction.

At the instant of conjunction if the *manaikyakhanda,* i.e. sum of the semi-diameters of the planets, is less than the algebraic difference of their latitudes there is no "*bheda yoga.*" What is of interest for further computations and observations is the *bheda yoga* – one planet "piercing" the other. This happens when the sum of the semi-diameters of the bodies is greater than the algebraic difference of their latitudes.

In case there is *bheda yoga*, the planet which is "lower" is treated as moon and the "higher" one as the sun. In that case the effects of parallax on the longitude and latitude namely *lambana* and *nati* are determined as in the case of a solar eclipse.

Note: Between two conjuncting bodies the *higher* and the *lower* between them are decided as follows:

[1] * See *Grahalaghavam'* English Exposition with Notes, etc. by S. Balachandra Rao and S.K. Uma, INSA, New Delhi, 2006

1. If both bodies have northern latitudes then the one with the higher latitude is considered as *higher*.
2. If both the bodies have southern latitudes then the one with numerically lesser latitude is considered as the *higher* body.
3. If the two bodies have their latitudes in opposite directions then the one with the northern latitude is the *higher* body.

Remark: While the conjunctions in longitudes of any two planets occur frequently, in the special sense of one "eclipsing" the other, such occurrences are extremely rare. This is so because the angular diameters of the planets are very small and therefore the sum of their semi-diameters is rarely greater than the difference between their latitudes.

However, the commentator of *Karanakutuhalam,* Sumatiharsa works out in detail the conjunction of *Guru* and *Sukra* which took place on *Salivahana Saka* 1541, *Vaisakha krsna* 14, Sunday. Though it was a case of a perfect conjunction (*bheda yoga*) it was not visible at the given place since the planets were below the horizon.

A very rare "close conjunction" (*bheda yoga*) of Jupiter and Venus occurred on January 4, 1818 at about 3^h 22^m a.m. (IST). According to our improved *siddhantic* procedure (ISP), the actual obscuration, geocentrically, was just for about 7^m44^s from $3^h17^m40^s$ a.m. to $3^h25^m24^s$ a.m.

1.12 Effects of Parallax – *Lambana* and *Nati*

In the computations of eclipses, the centre of the earth is taken as the reference point. However in reality, the observer is on the surface of the earth. It was noticed by ancient Indian astronomers for long that there was a marked departure in the observed time of conjunction of the sun and the moon from the computed positions. They realised that the distance between the centre of the earth and the position of the observer on the earth's surface was the cause of the difference between the *apparent* conjunction and the *computed* one.

It was also realized that the maximum value of the difference between these two timings was about four *ghatis,* i.e. 1 hour 36 minutes. However by modern computation it is known that this maximum value is slightly greater than four *ghatis*.

1.12.1 Parallax of the Moon

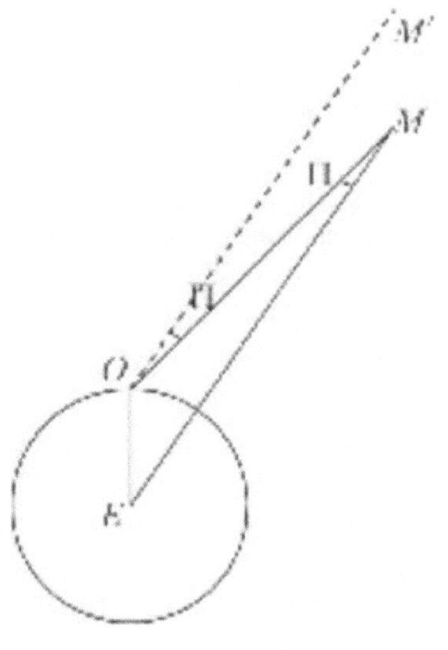

Fig. 1.2 Moon parallax

The parallax of the moon, called *lunar* parallax, for an observer at O on the earth's surface is explained through Figure 1.2. Suppose M is the centre of the moon and E that of the earth. For an observer at O on the earth's surface, the moon's centre M appears to be along the direction OM whereas from the earth's centre the direction is EM. The difference between these two directions is given by the angle $OME = \Pi$.

This angle is called the *parallax* of the moon for the observer at O. In fact, parallax is the angle subtended at M by the earth's radius at O. Since the earth is not perfectly spherical its radius is slightly different for observers at different places on the earth's surface. Hence, at a given time for the same position of M the parallax is slightly different for different places of observation on the earth.

As a heavenly body moves in the sky, for the observer at the same place on the earth's surface, the parallax ($O\hat{M}E$ in Figure 1.2 goes on changing. When the body is at the zenith, i.e. exactly at the top of the observer (along the line EO) parallax vanishes, i.e. $\Pi = 0$.

On the other hand when the heavenly body is along the observer's horizon, the parallax is *maximum*. This is shown in Figure 1.1 In this case triangle OME is *right angled* at O. We have $sin\Pi = \frac{EO}{EM} = \frac{r}{d}$. Here, r is the radius of the earth at O and d is the earth-moon distance, at the given time. Therefore parallax, $\Pi = sin^{-1}\left(\frac{r}{d}\right)$.
In Figure 1.2., let OM' be drawn parallel to EM. This gives the direction of M as seen from E. For the observer at O the *apparent* direction of M is OM. This means that the apparent position of M appears to have come *down*. This phenomenon is described by Bhaskara II as follows: "Since an observer elevated from the centre of the earth by the radius of the earth sees the moon deflected down from the geocentric line of conjunction, the parallax arises due to the radius of the earth."

yatah kvardhocchtrito drastacandram pasayati lambitam |
sadhyate kudalenato lambanam ca natistatha ||

Sl. 11, Ch. 9, *Goladhyaya*

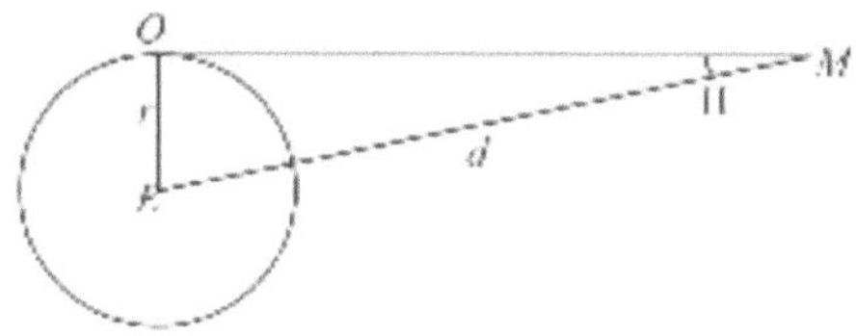

Fig. 1.3 Moon's horizontal parallax

1.12.2 Bhaskara II's Description of Parallax

Bhaskara II's description of the parallax is shown in Figure 1.4. O is the position of the observer on the surface of the earth centred at E. The positions of the sun and the moon at a *conjunction* with respect to E are represented by S and M, respectively.

The three points E, M and S are on the same line. For the observer at O, M and S are *apparently* not in conjunction, the directions OS and ES are called respectively *drkk-sutra* and *garbha-sutra* of the heavenly body S.

The angle OSE which is the parallax of the sun is also considered as the moon's parallax *relative* to the sun.
When both M and S are in the direction of the observer's zenith, i.e. in the direction of EOZ the parallax of both the sun and the moon vanishes. This means that there is no parallax for a body at the zenith. Bhaskara II, in this context, says "*khamadhye nasti lambanam.*" (*Sl.* 16, Ch. 9, *Goladhyaya*).

The ancient Indian astronomers took the maximum (i.e. horizontal) parallax as four *ghatis* in time unit and gave the expression for the parallax as four sin z where z is the zenith distance of the body. This corresponds to the modern expression, $p = \Pi sinz$ where Π is the horizontal parallax.

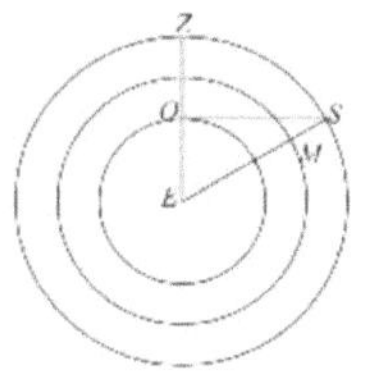

Fig. 1.4 Topocentric and geocentric lines of observation

1.13 Transits of Venus and Mercury in Indian Astronomy

We watched with great interest and curiosity the exciting phenomenon of **Venus transit** on June 8, 2004 on the TV as also its reflected image through the telescope. The recent **transit of Mercury** was on the night between 8 and 9 of November 2006.

The transits of Mercury and Venus occur when either of them is in conjunction with the sun as observed from the earth, subject to the prescribed constraints.

The transit of Venus is a less frequent phenomenon as compared to that of Mercury. For example, after the transit of Venus in June 2004 (Figure 1.5), the next occurrence will be in December 2012. After that, the subsequent Venus transit will be about 105.5 years later.

While detailed working of planetary conjunctions are discussed in all traditional Indian astronomical texts under the chapter *Grahayuti*, it has to be noted that the transits of Mercury and Venus are not *explicitly* mentioned. This is mainly because when either of these inferior planets is close to the sun it is said to be "combust" (*asta*) and hence not visible to the naked eye.

In a transit of Mercury or Venus the concerned tiny planet passes across the bright and wide disc of the sun as a small black dot.

In the following section we discuss the procedure, developed by us within the framework of *siddhantic* framework, for computing the circumstances of transit.

1.13.1 Improved Siddhantic Procedure of Transit Computations

We provide below a procedure, evolved by us, inspired by our traditional *siddhantas* and research papers of the late Prof. T. S. Kuppanna Sastry for computing a transit (Rao and Venugopal 2009). It is important to note that for a transit the inferior planet is always retrograde (*vakra*).

(a) The instant of conjunction of the true sun and the true planet (Mercury or Venus) is determined.

(b) At the instant of conjunction the following parameters are determined:

1. True positions of the sun and the planet.
2. True daily rates of motion of the two bodies (SDM and PDM).
3. The geocentric latitude of the planet β.
4. The horizontal parallax of the two bodies (PAR_1 and PAR_2).
5. Ascending node of the planet.
6. Angular diameters of the bodies (SDIA and PDIA).
   (Notes: Items (2), (3), (4) and (6) are in minutes of arc.)
7. $\mathrm{VRKNG} \equiv \frac{(\mathrm{PDM}-\mathrm{SDM})}{60}$, difference of rates of motion per *nadi*.
   (1 day = 60 *nadis* or *ghatis* = 24 hours)
   1 *nadi* $=\frac{2}{5}$ hour = 24 minutes.
   VRKNG: *Vyarka Graha Nadi Gati*
8. Let $\mathrm{PAR} = \mathrm{PAR}_1 + \mathrm{PAR}_2$, sum of the parallaxes in minutes of arc.
9. $\mathrm{D} = \mathrm{PAR}+\frac{(\mathrm{SDIA}+\mathrm{PDIA})}{2}$
   sum of the parallaxes and the semi-diameters.
10. $\mathrm{D}' = \mathrm{PAR}+\frac{(\mathrm{SDIA}-\mathrm{PDIA})}{2}$
    sum of the parallax with the difference of the semi-diameters.

11. Corrected latitude of the planet:

    $\lambda = \beta \times \frac{204}{205}$ in minute of arc.

12. Condition for the occurrence of the transit and its totality:

    (i) If $|\lambda| < \mathrm{D}$, the transit occurs

    (ii) If $|\lambda| < \mathrm{D}'$, the transit is total

    where VRKNG is the difference of rates of motion of the planet and the sun per *nadi*.

13. The middle of the transit is obtained by applying a small correction to the instant of conjunction. This correction is given by $\frac{-\lambda P}{(\dot{m})^2} + P^2$*nadi* where P = Rate of change in planet's latitude per *nadi*

14. Half-duration of the transit is given by

    $\mathrm{HDUR} = \frac{\sqrt{\mathrm{D}^2 - \lambda^2}}{\dot{\mathrm{m}}}$ in *nadis*,

15. Half-duration of the totality is given by

    $\mathrm{THDUR} = \frac{\sqrt{(\mathrm{D}')^2 - \lambda^2}}{\dot{\mathrm{m}}}$ in *nadis*,

16. Beginning of the transit (I contact, external ingress):

    BEGG = MIDDLE – HDUR

17. Beginning of the totality (II contact, internal ingress):

    BEGGT = MIDDLE – THDUR

18. Middle of the transit is the instant of greatest obscuration.

19. End of the totality (III contact, internal egress):

    ENDT = MIDDLE + THDUR

20. End of the transit (IV contact, external egress):

    END = MIDDLE + HDUR

Fig. 1.5 Venus transit

1.13.2 Summary of the Circumstances of Venus Transit of June 8, 2004

By following the procedure described in the earlier section, we get the local circumstances: (All timings are in IST)
Beginning of the transit BEGG = 10^h40^m (External ingress)
Beginning of totality BEGGT = 10^h59^m (Internal ingress)
Middle = 13^h52^m
End of the totality ENDT = 16^h44^m (Internal egress)
End of the transit END = 17^h03^m (External engress)
We note that according to modern computation, the instants of beginning, middle and the end are respectively 10^h45^m, 13^h51^m and 16^h58^m (IST).

1.14 *Rohini Sakata Bheda*

The *Rohini Sakata Bheda* (*RSB*) is an interesting configuration in which an eligible heavenly body enters the region marked by a group of stars including and very close to the *yogatara* (junction star) *Rohini* (Aldebaran). In Indian astronomical literature we find frequent references to *RSB*. The phrase *sakata bheda* means "breaking the cart" (of *Rohini*).

1.14.1 Stakeholders of RSB

The eligible stake-holders of the *RSB* phenomenon are the moon, Mars and Saturn. Among the traditional texts, it is the *Grahalaghavam (GL)* (Rao and Uma 2006) of Ganesa Daivajna (1520 AD) that is the most explicit in giving the details for *RSB* event, particularly by the moon. Ganesa prescribes the following:

gavi nagakulave khago'sya ced yamadigisuh khasaranguladhikah |
kabhasakatamasau bhinatyasrk sanirudupo yadi cejjanaksayah ||

"If a planet is at 17° of *Vrsabha* having its latitude southern and greater than 50 *agulas* (i.e. $150' = 2°30'$ south) it is said to break the *Rohini* cart (wain)......"

-*GL*, Ch. 11, *l.* 7

The texts mention the moon, Mars and Saturn as the possible stake-holders of *RSB*. Ganesa Daivajna further prescribes the condition for the moon to hit the *Rohini* cart:

svarbhanvaditibhato'sta rkasasamsthe sitasuh kabhasakatam sada bhinatti |
bhaumarkyoh sakatabhida yugantare syat sedanim na hi bhavatidrsi svapate ||

"As long as *Rahu* is in the eight *naksatras* from *Punarvasu* (up to *Citra*) the moon breaks the *Rohini* cart...."

-*GL*, Ch. 11, *Sl.* 8

He further points out that the *RSB* by Mars and Saturn take place after very long period (thousands of years) and that the event does not take place now. In fact for moon's *RSB* event, which take place more frequently, while the moon must be around *Nirayana* 47°, its node *Rahu* must lie between 80° and 186°40′.

1.14.2 Rationale for Moon's RSB

The *Rohini* asterism, according to Indian traditional texts, comprises five visible stars with southern latitudes in the constellation of *Vrsabha* (Taurus). The celestial (sidereal) longitudes of the stars in the above group range from 45°46′ to 49°45′ of which the central longitude is about 47°. The latitudes of these stars vary from 2°36′ to 5°47′ to the south of the ecliptic considering the measure of one *angula* as 3 arc-minutes, the above range comes to 52–116 *agulas*. Therefore for a body to enter the cart of *Rohini* it must have its southern latitude more than about 2°.5, i.e. 50 *agulas*. We have the traditional expression

$$southern\ latitude\ (sara)\ of\ the\ moon\ \beta := -270' sin(M-R) = 270' sin(R-M) \quad (1.12)$$

This give the range for moon's node *Rahu* from about 80°.4–193°.6, i.e. from the beginning of *Punarvasu* to the middle of *Svati*. However from the modern expression for the moon's latitude the leading term is $18461'' \sin F$ where $F = M - R$, the mean distance of the moon from its ascending node. Taking the sidereal longitude of the moon $M = 47°$, the range for sidereal *Rahu* is from 77°.47–196°.53.

The moon's sidereal period is about 27.32166 days. *Rahu's* sidereal period being 6793.47 days, it takes about 2246.74 days to cover the above prescribed range which is of length 119°.06. During this period of *Rahu's* stay in that range, the sidereal moon comes to 47° about 82 times. This means that in the course of *Rahu's* sidereal period of about 18.6 years, the maximum number of times the moon enters the *Rohini* cart is 82 times successively.

Now, out of *Rahu's* sidereal period of 6793.47 days (18.6 years) the remaining part during which the moon does not hit the *Rohini sakata* is about 12.45 years. However, the RSB by the moon with *Rahu* at a particular position, in the prescribed range repeats once in about 55.8 years.

1.14.3 Conjunction of Guru with Tisya Naksatra

In the *Rigveda Samhita* (*RV*), hailed as the most ancient literary work, and also in the *Atharva Veda* (*AV*) the rising of the effulgent Jupiter is described. Jupiter's conjunction with the *Tisya* (*Pusya*, i.e. δ Cancri) *naksatra* is specifically mentioned in the *Taittiriyasamhita* (*Yajurveda*):

> *brhspatih prathamam jayamano maho jyotisah parame vyoman* |
> *saptasyastu vijato ravena saptarasmiradhamat tamamsi* ||

> "*Brhaspati* when being born in the highest heaven of supreme light, seven-mouthed multi-form (combined) with sound and seven-rayed, has subdued the darkness."
>
> - *RV*, 4.50, 4: *AV*, 1088.4

There is a more specific reference to the location of Jupiter in the following *mantra* from the *Taittiriya Samhita*:

> *brhaspatih prathamam jayamanastisyam* |
> *naksatramabhisambabhuva* ||

> "*Brhaspati*, when first appearing, rose in front of the *Tisya* (*Pusya*) constellation."
>
> - *Tai. Sam.* 3.1.5

Again, the conjunction of *Guru* with *Pusya naksatra* (δ Cancri) being a rare event, it is worth investigating when this special conjunction took place so that on that basis the date of the *Rigveda* can be fixed. The latitude of *Pusya* is $0°004'37''$ north, very close to the ecliptic. The sidereal longitude is about $104°052'$. The inclination of Jupiter's orbit with the ecliptic being about $1°018'$, the said conjunction is feasible.

1.15 Indian Astronomical Tables

Indian astronomical almanacs (*pancangas*) provide the details of the five traditional calendrical items viz. *tithi, naksatra, yoga, karana* and *vara* (weekday) for each day in the course of a year. Further, true positions of the sun, the moon and planets are also provided. In addition to these, the days on which religious festivals like Ganesa Caturthi, Sri Rama Navami, etc. fall will also be specified.

For working out all these items, the *pancanga* – makers make use of the ready-to-use traditional astronomical tables called *sarinis*. These are compiled by medieval Indian astronomers based on classical texts like *Suryasiddhanta, Aryabhatiyam, Brahamasphutasiddhanta, Grahalaghavam,* etc.

Some of the well-known *sarinis* are (1) *Makaranda sarini*, (2) *Vakya paddhati*, (3) *Pratibhagi sarini*, (4) *Tithicintamani*, etc. These tables are being used even now for compiling traditional *pancangas*. There is a great need for critical studies in the these *sarinis*. While most of the basic classical texts were composed in ancient and early medieval periods, the astronomical tables (*sarinis*) were composed much later for computations of day-to-day calendrical and astronomical phenomena like eclipses. In composing these tables, the authors have incorporated corrections (*bijas*) and suitable changes based on their contemporary observations. Thus, the *sarinis* are also records of observations during the epochs chosen by the authors of the different *sarinis*.

David Pingree's conclusion, that the *bijasamskaras* (corrections) were incorporated in the *sarinis* of one system to bring them on par with the *sarinis* of another system, is too casual and frivolous if not biased and hazardous!

1.16 Conclusion

In the present chapter some glimpses of Indian astronomy are presented while some specific concepts and topics are highlighted. Still a lot of deeper studies in unpublished manuscripts, scattered over the entire country and outside, need to be taken up. The superiority of Indian astronomical procedures and parameters over the Ptolemaic and even the Copernican systems cannot be denied.

References

1. Apte DV (1942) *Tithicintamani* of Ganesa Daivajna with com. of Visvanatha - Ed Poona
2. Arkasomayaji D. *Siddhanta-siromai* of Bhaskara II – Eng. Exposition Kendriya Skt. Vidyapeetha (now RSVP), Tirupati, 1980; Reprinted 2000
3. Bhat Ramakrishna M, Banarsidass Motilal (1981) *Brhat Samhita* of Varahamihira, Eng. Tr. and Notes Delhi
4. Billard Roger (1971) *L'astronomie indienne:* . . . , publications de l'ecole francaise d'extreme orient, lxxxiii; Paris
5. Burgess E (1989) *Suryasiddhanta* – Tr., ed. and reprinted by Phanindralal Gangooli with Intn. by PC. Sengupta, Motilal Banarsidass Delhi
6. Chatterjee Bina *Khandaakhadyaka* of Brahmagupta – Ed. and Tr. with com. of Bhaottpala in 2 parts, New Delhi, 1970.
7. Dikshit SB (1969) *Bharatiya Jyotisa Sastra* (I Vol.) – Tr. by R.V.Vaidya, Govt. of India Pubn. Delhi
8. Dvivedi Sudhakara (1910) *Mahasiddhanta* of Aryabhata II – Ed. with Skt. com. Benares
9. Dvivedi Sudhakara (1929) *Siddhanta-siromani* of Bhaskara II – with auto-com. *Vasana*, Ed. Kashi Skt. Series No. 72 Benares
10. Ketkar Venkatesa (2006) *Ketaki-Grahaganitam* Bharatiya Vidya Prakasana Delhi
11. Ketkar Venkatesa (2008) *Jyotirganitam* Chowkhamba Krishnadas Academy Varanasi
12. Kern H (1990) *The Aryabhatiya* (of Aryabhata) with com. Bhatadipika of Paramadisvara (2nd ed.)–Ed. Eastern Book Linkers Delhi
13. Misra Babuaji (1932) *Siddhanta sekhara* of Sripati – Ed. with com. of Makkibhatta (for Ch.I-IV) and an original com. (Ch. IV-X) Calcutta Uni. Press Calcutta
14. Ramasubramanian K, Srinivas MD, and Sriram, MS (May 1994) "Modification of Earlier Indian Planetary Theory by the Kerala astronomers and the Implied Heliocentric Picture of Planetary Motion" *Current Science*
15. Rao Balachandra S (2005) *Indian Mathematics and Astronomy – Some Landmarks* –(Rev. 3rd Ed.), Bhavan's Gandhi Centre of Science and Human Values Bangalore
16. Rao Balachandra S and Uma SK (2007-08) *Karana-kutuhalam* of Bhaskara II – with Eng. Tr., Notes, Examples etc. JHS Vol 42 Nos. 1-2 & Vol. 43 Nos. 1& 3 INSA New Delhi
17. Rao Balachandra and Venugopal Padmaja (2008) *Eclipses in Indian Astronomy* – Gandhi Centre of Science and Human Values Race Course Road Bangalore
18. Rao Balachandra S and Venugopal Padmaja (2009) *Transits and Occultations in Indian Astronomy* – Gandhi Centre of Science and Human Values Race Course Road Bangalore
19. Rao Balachandra S. and Uma SK (2006) *Grahalaghavam* of Ganesa Daivajna – with Eng. Exposition, Notes examples etc. IJHS INSA New Delhi
20. Rao Balachandra S, Uma SK, and Venugopal Padmaja (2008) *Conjunctions, Transits and Occultations by Siddhantic Procedures* IJHS Vol. 43 Now 4 p.515-551 INSA New Delhi
21. Rao Balachandra S, Uma SK, and Venugopal Padmaja (2003) *Lunar Eclipse Computation in Indian Astronomy with Speical Reference to Grahalaghavam* IJHS Vol. 38 No.3 p.255-271 INSA New Delhi
22. Rao Balachandra S, Uma SK, and Venugopal Padmaja (2004) *Mean Planetary Positions According to Grahalaghavam,* IJHS Vo. 39 No.4 p. 441-466 INSA New Delhi
23. Sarma KV (1936) *Drgganitam* of Paramesvara – Cr. Ed. with Intn. VVRI Hoshiarpur
24. Sarma KV (1977) *Tantra-sangraha* of Nilakantha Somayaji with *Yuktidipika* and *Laghuvivrti* com. of Sankara – Cr. Ed. with Int. and App. VVBIS & IS Hoshiarpur
25. Sarma KV (2008) *Ganita-Yukti-Bhasa* of Jyesthadeva, Vols. I & II, Malayalam text – Cr. Ed. with Eng. Tr. with Explanatory Notes in Eng. By K Ramasubramanian, MD Srinivas and MS Sriram, Hindustan Book Agency New Delhi

26. Sarma KV (2001) *Sadratnamala* of Sankaravarman, - Cr. Ed. IJHS Vol. 36 Nos. 3-4 (Supplement) INSA New Delhi
27. Sastri Sambasiva K (1977) *Aryabhatiyam* of Aryabhata with com. by Nilakantha Somasutvan, 3 Vols; Ed. Uni. of Kerala Trivandrum
28. Sastry, Kuppanna TS (1985) *Vedanga Jyotisa of Lagadha* – Cr. Ed. By K.V.Sarma INSA New Delhi
29. Sastry, Kuppanna TS (1993) *Pancasiddhantika* of Varahamihira – with Eng. Tr. and Notes Cr. Ed. by Dr. KV Sarma PPST Foundation Madras
30. Sastry Kuppanna TS (1989) *Collected Papers on Jyotisa*, Kendriya Skt. Vidya Peetha (now RSVP) Tirupati
31. Sengupta PC (1934) *Khanda-khadyaka* of Brahmagupta – Ed. and Tr. with com. of Caturveda Prathudakasvami Calcutta
32. Sharma Rama Swarup (1966) *Brahmasphuta siddhanta* of Brahmagupta, Ed. with Vasana com. 4 Vols. Ind. Inst. of Astron. and Skt. Res. New Delhi
33. Shukla KS and Sarma KV (1976) *Aryabhatiya* of Aryabhata I – Cr. Ed. with Eng. Tr. and Notes INSA New Delhi
34. Shukla KS (1996) (Ed.) *Aryabhatiya* of Aryabhata with com. of Bhaskara I and Somesvara INSA New Delhi
35. Shukla KS (1960) *Maha-Bhaskariya* of Bhaskara I with Eng. Tr. and Notes Lucknow University Lucknow
36. Shukla KS (1960) *Laghu-Bhaskariya* of Bhaskara I with Eng. Tr. and Notes Lucknow Universty Lucknow
37. Shukla KS (1985, 1986) *Vatesvara siddhanta* and *Gola* of Vateavara – Cr. Ed. with Eng. Tr. and com. Parts I & II INSA New Delhi
38. Shukla KS (1990) *Laghumanasam* of Manjula – Cr. Study Tr. and Notes IJHS Vol. 25 Nos. 1-4 (Supplement) INSA New Delhi
39. Shukla KS (1957) *Suryasiddhanta* with com. of Paramesvara – Ed. Lucknow Uni. Lucknow
40. Sriram MS, Ramasubramanian K and Srinivas MD (2002)*500 years of Tantrasangraha* – Ed. IIAS Shimla
41. Upadhyaya AK (1996, 1998) *Siddhanta-darpanah* of Samanta Candrasekhara Simha – with Hindi Tr. and Eng. Tr. Nag Publishers Delhi
42. Waerden Van der BL (1980) (Zurich Uni.), *Two Treatises on Indian Astronomy* JHA xi

Chapter 2
Astronomy in *Gaṇita-Yuktibhāṣā* of Jyeṣṭhadeva (c. 1530)

M. S. Sriram

2.1 Introduction

Kerala had a continuous and vigorous tradition of astronomy and mathematics, at least from the 7th century (Sarma 1972). What can be called the 'Kerala school of astronomy and mathematics' evolved here during 14th–17th centuries. Mādhava of Saṅgamagrāma (1340–1425) was the father figure of this school. All the later astronomers of Kerala invariably attribute the path-breaking results in infinite series for the inverse tangent, sine and cosine functions and many other innovations in astronomical calculations to him.

Jyeṣṭhadeva's *Gaṇita-Yuktibhāṣā* (rationales of mathematical astronomy), popularly known as *Yuktibhāṣā* is a seminal text of the Kerala school (Sarma et al. 2009). *Yuktibhāṣā* is composed in Malayalam, which is the language of Kerala. This is perhaps the first instance of significant original work in Mathematics and Astronomy which is written in a regional, spoken language. *Yuktibhāṣā* is exclusively devoted to explanations, proofs and demonstrations of results in mathematics and astromomy known to Indians at that time. It is hard to find an equivalent for the word *Yukti* as employed in Sanskrit or Malayalam languages. It stands for 'rationale' or 'proof' or '*upapatti*' or 'reasoning' or 'devised strategy' or 'demonstration of'. *Bhāṣā* is the spoken language of a region. Hence the title of the text *Yuktibhāṣā* means "rationales explained in the local language" which happens to be Malayalam in this case. A Sanskrit version of the text is also available (Sarma 2004). However it is clear that it is a rough and ready translation into Sanskrit of the Malayalam original. *Yuktibhāṣā* is in prose and the date of composition has been assigned to be around AD 1530.

The importance of the work was brought to the notice of modern scholarship for the first time by C. M. Whish in the 1830s (Whish 1834). However it did not catch the attention of scholars till the 1940s when K. Mukunda Marar, C. T. Rajagopal, A. Venkataraman and others wrote many papers on the proofs of infinite series for π and the sine and cosine functions contained in *Yuktibhāṣā* and other Kerala works[1]. The mathematics part of *Yuktibhāṣā* is explained in modern Malayalam in an edition by Ramavarma (Maru) Thampuran and Akhileswara Ayyar in 1948 (Tampuran and Akhileswar Ayyar 1948). Now we have a complete translation of both the parts of *Yuktibhāṣā* (mathematics and astronomy) into English with detailed explanatory notes (Sarma et al. 2009).

There have been a large number of articles highlighting the significant results in the mathematics part of *Yuktibhāṣā*. [2] (Divakaran 2007), and their importance in the history of mathematics is now appreciated to some extent. However, there seem to have been no study of the astronomy part of *Yuktibhāṣā*, except for the overview given by K. V. Sarma and S. Hariharan (Sarma and Hariharan 1991). This chapter is an attempt in that direction.

Yuktibhāṣā is part of a long tradition of mathematical astronomy in India, beginning with *Āryabhaṭīya* (c. 499) of Āryabhaṭa, which set the framework for later works on astronomy in India. In Section 2.2, we discuss the salient features of *Āryabhaṭīya* and other major texts of Indian astronomy. In Section 2.3, we provide a brief outline of the Kerala school of astronomy, with a special focus on *Tantrasaṅgraha* of Nīlakaṇṭha Somayājī, as it is the stated aim of *Yuktibhāṣā* to explain the techniques and theories employed in the former text. Among ancient Indian works, *Yuktibhāṣā* is unique in the manner in which mathematics and astronomy

Department of Theoretical Physics, University of Madras, Guindy Campus, Chennai-600 025, India.

[1] See the references listed in the prologue to Vol. I of *Gaṇita-Yuktibhāṣā* in reference 2

[2] See the references listed in the prologue to Vol. I of *Gaṇita-Yuktibhāṣā* in reference 2

are systematically developed. In Section 2.4, we describe the nature of *Yuktibhāṣā* and its contents, especially of its astronomy part. In Sections 2.5, 2.6 and 2.7, we highlight some important topics discussed in *Yuktibhāṣā*. Nīlakaṇṭha's model presented in his *Tantrasaṅgraha* and discussed in detail in *Yuktibhāṣā* comes closest to the Kepler model (which was to come more than a century later), as far as planetary positions in a geocentric framework are considered. The planetary model in *Yuktibhāṣā* is described in Section 2.5. For the sake of comparison, the geocentric position of a planet in Kepler's model is discussed in the Appendix. We have a very systematic and exact treatment of spherical astronomy problems and *dṛkkarma* or parallax correction in *Yuktibhāṣā*. As an example, we discuss the elegant derivation of the expression for the inclination of moon's orbit with equator in Section 2.6. As another example of the geometrical methods in *Yuktibhāṣā*, we briefly discuss the computation of *bimbāntara*, which is the distance between the centres of the solar and lunar discs in Section 2.7. In Section 2.8, we make a few concluding remarks.

2.2 Antecedents of *Yuktibhāṣā*: I. *Āryabhaṭīya* and its Aftermath

The *Siddhānta* period begins with *Āryabhaṭīya* which was composed in AD 499 (Sen 1971; Sen and Shukla 1985; Balachandra Rao 2000; Subbarayappa 2008). This refers only to the available texts; surely there were earlier *Siddhānta*s which are quoted by the later astronomers. Mathematics developed with astronomy in India. *Siddhānta*s contain correct mathematical solutions to astronomical problems, namely calculation of positions of planets (Sun, Moon, Mercury, Venus, Mars, Jupiter and Saturn), diurnal problems involving spherical astronomy, and eclipses. The diurnal problems would include, finding north-south directions, latitude of a place, sun's diurnal path, its declination, sunrise/sunset times, measurement of time (from shadow), relations among various celestial coordinates, calculation of *lagna* (point on the ecliptic which is on the horizon at any time), etc. The treatment of eclipses would include the calculation of the instant of conjunction, the two half-durations (from the first contact to the middle, and from the middle to the last contact), duration of totality, magnitude of the eclipse, and so on.

Āryabhaṭīya is the earliest available *Siddhānta* text in India, which contains a systematic acount of most of the traditional astronomical problems (Shukla and Sharma 1976). It is mentioned in the text itself that it was composed 3,600 years after the begining of *Kaliyuga*. This corresponds to AD 499. Further it is stated that the author was 23 at that time. Āryabhaṭa composed this work in *Kusumapura*, which is the same as Paṭalīputra (essentially modern Patna).

From *Āryabhaṭīya* onwards, one considers a *Mahāyuga* of 43,20,000 years. The number of revolutions in the stellar background made by the planets in the *Mahāyuga* are given in the texts (see Table 2.1). For Mercury and Venus, these are mentioned as the revolution numbers of their *śīghroccas*, which are nothing but their heliocentric revolution numbers. The number of civil days in a *Mahāyuga* known as the *Yugasāvanadina* is also specified. In *Āryabhaṭīya*, its value is 1,57,79,17,500.

Table 2.1 Planetary revolutions in a *Mahāyuga* as given in *Āryabhaṭīya*, and the inferred sidereal periods

Planet	No. of revolutions in a *Mahāyuga*	Period of revolution in days (sidereal period)	Modern value
Sun	43,20,000	365.25868	365.25636
Moon	5,77,53,336	27.32167	27.32166
Moon's apogee	4,88,219	3231.98708	3232.37543
Moon's nodes	2,32,226	6794.74951	6793.39108
Mercury	1,79,37,020	87.96988	87.96930
Venus	70,22,388	224.69814	224.70080
Mars	22,96,824	686.99974	686.97970
Jupiter	3,64,224	4332.27217	4332.58870
Saturn	1,46,564	10766.06465	10759.20100

From this, the mean longitudes of the planets can be calculated at any time. Normally, it is assumed that the mean longitudes are zero at the beginning of *Kaliyuga*. In *Āryabhaṭīya*, this is taken to be the mean sunrise at Ujjain of February 18, BC 3102.

Now, the apparent motion of the sun, moon and planets in the background of stars is not uniform. The planets (including the earth) and the moon move in elliptical orbits around the sun, and the earth, respectively. The ellipticity of the orbit takes the non-uniform nature of the motion into account. This is essentially Kepler's picture. In the geocentric frame of reference, the sun moves around the earth, and the geocentric longitude of the sun would be the heliocentric longitude of the earth increased by 180°. The geocentric longitudes of the planets would be related to their heliocentric longitudes and the geocentric longitude of the sun.

Even in the ancient times (before Kepler), the predictions of planetary positions were fairly accurate. This means that the ancient models would have been equivalent to the Kepler model in the geocentric frame, at least roughly (Sriram et al. 2002). In the Indian models, two corrections were applied to the mean planet to obtain the 'true' geocentric longitudes. These are:

(1) *Manda-saṃskāra*: This is due to the non-uniformity of motion resulting from the eccentricity of the planet's orbit. An epicyclic or eccentric model is used to obtain the *manda*-correction. This corresponds to the 'equation of centre' in modern terminology. The *manda*-corrected mean planet is called *mandasphuṭa-graha* or simply *mandasphuṭa*. This is the true geocentric longitude in the case of the sun, and essentially so for the moon (for which some other minor corrections are specified in later texts). In the case of the actual planets called *tarāgraha*s (traditionally, only Mercury, Venus, Mars, Jupiter and Saturn), the *mandasphuṭa* is the true heliocentric longitude.

(2) *Śīghra-saṃskāra*: For *tarāgraha*s, one more correction, namely *śīghra*, has to be applied to obtain the true longitude called *sphuṭagraha*. Hence, *śīghra-samskāra* converts the heliocentric longitudes of the *tarāgrahas* to geocentric longitudes.

It is in *Āryabhaṭīya* that the above two corrections are discussed clearly for the first time in India. This planetary model described by Āryabhaṭa roughly amounts to the planets orbiting around the sun in eccentric orbits, with the sun itself orbiting around the earth, though Āryabhaṭa does not state it as such. We discuss the Indian planetary models, including an error in the formulation for the interior planets, later. Significantly, the picture of latitudes of planets is broadly correct in *Āryabhaṭīya*.

We also list, for the sake of completeness, some of the later astronomers and their major works (Sen 1971; Sen and Shukla 1985; Rao 2000; Subbarayappa 2008): Varāhamihira (b. AD 505) (*Pañcasiddhāntikā*), Bhāskara I (7th century) (*Mahābhāskarīya*, *Āryabhaṭīyabhāṣya*), Brahmagupta (7th century) (*Brāhmasphuṭ siddh-ānta*), Vaṭeśvara (9th century) (*Vaṭeśvarasiddhānta*), Mañjulācārya (10th century) (*Laghumānasa*), Śrīpati (11th century) (*Siddhāntaśekhara*), Bhāskarācārya II (b. AD 1114) (*Siddhāntaśiromaṇi* with *Vāsanāb hāṣya*). *Siddhāntasiromaṇi* contains most of the results known to Indians at that time. It was commonly believed that Bhāskara II was the last great figure in the Indian tradition of mathematics and astronomy, and nothing of any significance happened afterwards, except for improvement of parameters in astronomy. Now it is recognised that there was very significant work in astronomy and mathematics in Kerala during 14th–17th centuries. We will discuss this in the next section.

Āryabhaṭīya has only 121 verses and is very cryptic. Later texts are more elaborate, but are still concise and are mainly algorthmic. It would be difficult to understand the contents by studying the texts alone. Fortunately we have commentaries on the major texts. Bhāskara I, Someśvara, Sūryadeva Yajvan, Parameśvara, Nīlakaṇṭha Somayājī and others have written commentaries on *Āryabhaṭīya*. Pṛthūdakasvāmin has written a commentary on *Brāhmasphuṭasiddhānta* of Brahmagupta. Bhāskara II has written a commentary, *Vāsanābhāṣya* on his own *Siddhāntaśiromaṇi*, and so on. The explanations in the commentaries are sometimes concise, or just clarifications. At times, they are very elaborate and go far beyond the text.

2.3 Antecedents of *Yuktibhāṣā*: II. *Tantrasaṅgraha* of Nīlakaṇṭha

2.3.1 Kerala School of Astronomy

Āryabhaṭīya became popular in Kerala soon after its composition. The astronomical parameters in *Āryabhaṭīya* were revised here and the revised system called *Parahitagaṇita* was enunciated by Haridatta in his *Grahacāranibandhana*. *Laghubhāskarīya* and *Mahābhāskarīya* of Bhāskara I which expounded the Āryabhaṭan school were popular here, as also *Laghumānasa* of Mañjulācārya. Mādhava (1340–1425) ushered in the golden period of the Kerala school. His known works like *Veṇvāroha*, *Sphuṭacandrāpti* and *Agaṇitagrahacāra* may not be major works conceptually. However, the distinctive contributions of the Kerala school in mathematical analysis, as well as some significant iterative methods in planetary computations are invariably attributed to him by the later astronomers from Kerala. Parameśvara of Vaṭasseri (c. 1360–1455), a student of Mādhava was a prolific writer, authoring about 30 works. Emphasizing the need for revising the planetary parameters through observations, he thoroughly revised the *Parahita* system and introduced the *Dṛggaṇita* system. Apart from *Dṛggaṇita*, some of his other major works are *Bhaṭadīpikā* which is a commentary on *Āryabhaṭīya*, *Goladīpikā*, *Siddhāntadīpikā*, and *Grahaṇamaṇḍana* on eclipses. He was one of the first astronomers to discuss the geometrical model of planetary motion in some detail.

Dāmodara was the son and disciple of Parameśvara. Little is known about his works. Nīlakaṇṭha Somayājī or Somasutvan (1444–1550) was Dāmodara's disciple. He hailed from Trikkantiyur near Tirur in south Malabar. *Tantrasaṅgraha* composed in AD 1500 was his major work (Pillai 1958; Sarma 1977; Sarma and Narasimhan 1998; Ramasubramanian and Siram, in press). We will discuss this text in some detail, shortly.

Apart from *Tantrasaṅgraha*, Nīlakaṇṭha composed many other works. *Āryabhaṭīya-bhāṣya* composed by him late in his life is perhaps the most elaborate commentary on *Āryabhaṭīya*, and is yet to be translated and studied in detail. *Golasāra* is a short work in 56 verses containing many details not covered in *Tantrasaṅgraha*. The importance of *Siddhāntadarpaṇa* lies in the fact that the author presents herein the astronomical constants as verified through his own observations and investigations. *Jyotirmīmāmsā* of Nīlakaṇṭha has a unique place in the history of Indian astronomy, as it exclusively focusses on epistemological issues concerning the science of astronomy and mathematics (Srinivas 1990). It strongly emphasises the role of observations and experimentation in revising astronomical parameters and settling any dispute.

Jyeṣṭhadeva of *Parakroḍa* or *Paraṅṅoṭṭu* family is the author of *Gaṇita-Yuktibhāṣā*. He was a pupil of Dāmodara initially, and probably recieved instruction from Nīlakaṇṭha later. Śaṅkara Vāriyar of Tirukkutaveli (c.1500–1560) was a disciple of Nīlakaṇṭha Somayājī and was also deeply influenced by Jyeṣṭhadeva. He is the author of two commentaries in Sanskrit on *Tantrasaṅgraha*, namely *Laghuvivṛtti* in prose, and a far more elaborate one *Yuktidīpikā* in verse. He also seems to have written a Malayalam commentary *Kriyākalāpa* on the same, and also the commentary, *Kriyākramakarī* on *Līlāvatī* of Bhāskara II. There are similarities in the treatment of various topics in *Yuktidīpikā* and *Yuktibhāṣā*.

Citrabhānu (c.1475–1550) the author of *Karaṇāmṛta* was also a disciple of Nīlakaṇṭha, whereas Acyuta Piṣāraṭi (c.1550–1621) of Trikkantiyur was a student of Jyeṣṭhadeva. Acyuta was the author of *Sphuṭanirṇayat antra* and *Rāśigolasphuṭanīti*. Putumana Somayāji's (c.1660–1740) *Karaṇapaddhati* is also considered an important work. *Sadratnamāla* of Prince Śaṅkara Varman (1800–1838) is a compendium of the Kerala school of mathematics and astronomy.

Tantrasaṅgraha

As mentioned earlier, the astronomy part of *Yuktibhāṣā* is patterned after *Tantrasaṅgraha*, and most of the algorithms in the latter are proved in *Yuktibhāṣā*. This is the reason for elaborating on the topics covered in *Tantrasaṅgraha*, here. It is a comprehensive text in the sense that all the procedures needed for the computation of quantities of physical interest such as longitudes and latitudes of planets, various diurnal problems, determination of time, eclipses, visibility of planets, etc., are discussed. However, explanations are not provided save on some occasions, as this belongs to the *Tantra* class of texts which are intended to be algorithmic in nature. Apart from *Yuktibhāṣā*, the explanations of the algorithms are also to be found in the two commentaries of the text, namely *Laghuvivṛtti* and *Yuktidīpikā*.

The revolution numbers of the sun, moon and the five planets given in this text differ little from the Āryabhaṭan values summarised in Table 2.1. The *Yugasāvanadina* is also the same as in *Āryabhaṭīya*. However, it is noteworthy that while specifying the number of revolutions of an inner planet, the word *svaparyayāḥ* is used, which clearly means that the revolution number given refers to its own revolution, and not to its *śīghroccas*, as specified in the earlier texts. This is of great significance in the computation of longitudes of the inner planets, and will be explained later.

The revolution numbers of the apsides and nodes of the planets which are needed for calculating the true longitudes are also given. The mean longitudes at the begining of *Kaliyuga* are not assumed to be zero, as in *Āryabhaṭīya*, and their values are specified. The mean longitude of a planet can be calculated at any given time, given these epochal values, the revolution numbers and the *Yugasāvanadina*.

For Mercury and Venus, the equation-of-centre-correction was wrongly applied to the mean sun, instead of the mean heliocentric planet, in the earlier Indian texts. This is true of Ptolemy's *Almagest* also, and the correction does not correspond to anything physical. It was Nīlakaṇṭha who set this right in *Tantrasaṅgraha*. Here the *manda* correction is applied to the mean heliocentric planet for the interior planets, just as in the case of exterior planets. This departure, as well as a clear analysis of the latitudinal motion, led him to propose a geometrical picture of planetary motion in his other works. According to this, the planets move in eccentric orbits around the mean sun, which itself moves around the earth (Ramasubramnian et al. 1994).

The diurnal problems (mostly related to the motion of the sun and the shadow cast by it) are dealt with exhaustively in the text. It is far more systematic than earlier Indian texts, and the formulae of spherical trigonometry stated here are exact. For instance, *lagna* is the longitude of the point on the ecliptic which is on horizon, at any given time. This is an important concept in Indian astronomy. An exact formula for the *lagna* is given in the text, perhaps for the first time. Another noteworthy feature is the inclusion of the effects of the finite size of the solar disc and the solar parallax, in the determination of the latitude of a place through the measurement of the noon shadow.

The actual distances of the sun and the moon from the centre of the earth play an important role in eclipse calculations, as these determine the apparent sizes of the solar and lunar discs, as well as their parallaxes. The mean distances are specified in the text. Two corrections are applied to these, for obtaining the actual distances. The first one is due to the eccentricities of the orbits. The second correction is essentially the 'evection' term, in the case of the moon. It may be recalled that the evection term is an important contribution of Ptolemy. It makes its first appearance in India in *Laghumānasa* of Mañjulācārya, which has been referred to by Nīlakaṇṭha. There is a similar correction for the sun, the significance of which is not clear. The true distance, which is called *dvitīya-sphuṭa-yojana-karṇa* is used in all relevant calculations.

A solar eclipse is very sensitive to the parallaxes of the sun and the moon, especially the later, as its magnitude is significant. The parallax problem is treated in far more detail in *Tantrasaṅgraha* than in earlier texts. The effect of parallax is taken into account in determining the exact instant of conjunction, half-durations of the eclipse, and its visibility.

The angle of separation between the centres of the solar and lunar discs determines the lunar phase, and is calculated exactly here. This is for an observer on the surface of the earth, and includes the effect of parallax. The procedure involves the distance of separation between two points in a three dimensional space, and is a precursor of coordinate geometry calculations.

There are other novel features in *Tantrasaṅgraha*, such as the alternate method to find the declination of the moon, involving the instantaneous angle of inclination between the lunar orbit and the celestial equator.

2.4 Nature of *Yuktibhāṣā*, and its Contents

The entire text of *Yuktibhāṣā* occurs as one continuum without any marking of subjects or division into parts. However, towards the middle of the work, there is a benedictory statement when the discussion of mathematics ends, and that of astronomy commences. Hence there is a natural division of the text into mathematics and astronomy portions. The mathematics part was divided into seven chapters by Ramavarma Tampuran and Akhileswara Ayyar. The astronomy part has been editorially divided into eight chapters by K.V. Sarma. A substantial part of mathematics known to Indians at that time is included and explained in the work. The astronomy part is very comprehensive, and includes almost all that was known in India then, save some topics like instruments.

Yuktibhāṣā is really a textbook for students. It systematically develops the subject and is self-contained. The basics are set forth at the beginning of each topic and all the results which follow are proved. The main concepts are often repeated for emphasis. Alternate methods for important formulae are given, wherever possible. There are no equations as we write now, nor do we find any diagrams in the available manuscripts. However, from the detailed descriptions given in the text, one can write down the equations, as also draw the figures without ambiguity. In the very begining of the work, Jyeṣṭhadeva says:

> Having bowed at the most venerable feet of the teacher, the entire calculation, whatever is needed for the computation of the motions of the planets is being set out by me.

Then he says:

> Here at the outset, with a view to expound, following *Tantrasaṅgraha*, all the calculations .....

This is true only in a general sense, and the mathematics part is actually an independent treatise. The nature of numbers and various mathematical operations like addition, multiplcation, division, squaring and square root are discussed in the first two chapters. The third chapter is on fractions, including various ingenious manipulations with them. The fourth chapter is on the rule of three, where both general and inverse rules are set out. The fifth chapter is on linear indeterminate equations or *kuṭṭākāra* with specific applications to problems in astronomy. The famous results on the infinite series for π including fast convergent versions of them, and their proofs form the subject matter of the sixth chapter. The infinte series for the inverse tangent function is also included.

Sine and cosine functions appear everywhere in positional astronomy. In fact, it is in the context of astronomy that trigonometry was invented. In most Indian texts, invariably there would be verses describing the sine table comprising the sines of the first 24 multiples of $225'$ (that is, $3^\circ\ 45'$ to 90° in steps of $3^\circ\ 45'$). The sines of the intermediate values are to be found by interpolation. The seventh chapter of *Yuktibhāṣā* begins with a description of the method to generate the above kind of sine table. It goes on to explore more accurate determination of sine and cosine functions for an arbitrary angle, and it is in this context that the infinite series for them are described. The infinite series are based on the fact that the differentials of $\sin\theta(\cos\theta)$ are proportional to $\cos\theta(\sin\theta)$. The accurate value of sines and cosines and their differentials are meant for use in astronomical computations. The seventh chapter also includes a derivation of the surface area and volume of a sphere using methods which amount to carrying out integration, and also of the area and diagonals of a cyclic quadrilateral.

Chapters 8–15 in *Yuktibhāṣā* are on astronomy. The eighth chapter is on planetary theory. The epicyclic and eccentric theories are described for both *mandasamskāra*(equation of centre) and *śīghra-samskāra* (conversion from heliocentric to geocentric coordinates). The correct formulation of the equation of centre for interior planets is incorporated in the planetary theory, and the difference with the earlier formulations is noted. The geometrical model of Nīlakaṇṭha is described in detail. The computation of geocentric latitudes is also described. The orbit of a planet has to be projected onto the plane of the ecliptic to obtain the longitude. This is explained. There is an interesting discussion on how one can obtain the geocentric coordinates of a celestial body, given the luni-centric coordinates. This indicates how comfortable Jyeṣṭhadeva was with coordinate transformations while going from one frame of reference to another, in the context of astronomy.

The ninth chapter deals with the celestial sphere and the related great circles such as meridian, horizon, equator, ecliptic and their secondaries, diurnal circles, and so on. *Bhūgola* is the terrestrial sphere, whereas *Vāyugola* and *Bhagola* refer to the celestial sphere with the celestial equator and the ecliptic as the principal circles, respectively. The treatment of the topics is very similar to that in a typical modern textbook on spherical astronomy. One of the principal results in this chapter is the expression for the declination and right ascension of a planet with latitude.

As mentioned earlier, *Yuktibhāṣā* contains a very systematic exposition of problems of spherical trigonometry. One can define six observer-independent celestial quantities related to the sun, like longitude, right ascension, declination, etc. Given two of them, the other four can be detemined in terms of the known two. This can be done in 15 different ways. The tenth chapter deals with these 15 problems. This is a prelude to the '10 problems' in the next chapter. The 15 problems are not found in *Tantrasaṅgraha*.

The eleventh chapter deals with directions and gnomic shadow. This chapter is very elaborate and comprehensive, following closely the corresponding chapter in *Tantrasaṅgraha*, except that *Yuktibhāṣā* also presents a systematic derivation of all the results. The chapter begins with the method for finding accurately the east-west and north-south lines from sun's shadow at different times. This includes the small correction due to the change in declination of the sun over a day. In diurnal problems, the latitude of the observer plays an important role. This is found from the midday shadow of the sun. Here two corrections due to the finite size of the sun, and its parallax in the determination of the latitude are included. Another important problem discussed in *Tantrasaṅgraha* that is explained in this chapter is of finding the time from the gnomic shadow.

One of the important topics discussed in *Tantrasaṅgraha*, is *Daśapraśnāḥ* or the ' 10 problems', the rationale for which are to be found in *Yuktibhāṣā*. Consider the five quantities, the zenith distance of the sun(z), its hour angle (H), its declination (δ), its Azimuth (A) and the latitude of the place (ϕ), as indicated in Figure 2.1. Now, there are 10 different ways to choose any two out of the five. The subject matter of *Daśapraśnāḥ* is to determine any two of them, given the other three. Perhaps it is for the first time that a problem of this type is posed and systematically solved. The solutions correspond to exact spherical trigonometrical results.

The parallax corrections to the latitude and longitude known as *lambana* and *nati* which are crucial in eclipse calculations are described. These are exact and do not assume that moon's latitude is small during an eclipse. The eleventh chapter also includes the procedure for finding the *lagna* and the *kālalagna* (time elapsed after the rise of the vernal equinox over the horizon).

The twelfth chapter is on eclipses. As most of the relevant quantities had already been discussed in the previous chapters, this chapter is a short one. Procedures for the computations of the half-durations, eclipsed portion at required time, etc. are given. A graphical description of the evolution of an eclipse is also provided in this chapter.

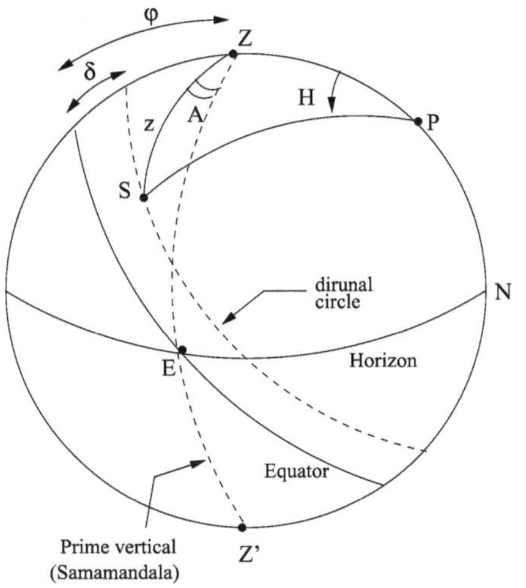

Fig. 2.1 The five quantities involved in the set of problems dealing with *Daśapraśnāḥ*

The thirteenth chapter is on *Vyatīpāta*, which is a concept peculiar to Indian astronomy. This occurs when the magnitudes of the declinations of the sun and moon are equal, whereas their rates of change have opposite signs. Here an alternate formula for the declination of the moon which is given in *Tantrasaṅgraha* is explained using a very elegant method.

Visibility correction for the planets is discussed in the fourteenth chapter. The method for finding the *lagna* corresponding to the instant at which a planet with latitude rises in the eastern horizon is also described.

'Elevation of the moon's cusps' is the subject matter of the fifteenth chapter. This is essentially the angle between the line joining the tips of the lunar crescent and the horizon. This involves the *bimbāntara* which is the distance between the centres of the lunar and solar discs on the celestial sphere, including the effect of parallax. The method given here is a clear forerunner of the techniques of three dimensional coordinate geometry. The text in all the available manuscripts ends abruptly without giving the method to find the elevation of the lunar cusps using *bimbāntara*.

2.5 Planetary Model in *Yuktibhāṣā*

2.5.1 Correction due to eccentricity of the orbit in Indian planetary models Manda-samskāra

We consider the explicit geometrical construction for computing the correction due to eccentricity described in *Yuktibhāṣā*. Essentially the same form of the correction term has been described right from *Āryabhaṭīya* down to *Tantrasaṅgraha*. The uniqueness of *Yuktibhāṣā* is in the detailed geometrical description.

In Figure 2.2 the dashed circle with O as the centre and radius R is the *kakṣyāvṛtta* or the 'deferent circle'. In the Indian texts, the distances in geometrical figures associated with planetary motion are expressed in minutes. Normally, R is the *trijyā* which is $3438'$ (number of minutes in a radian). A represents the direction of the *mandocca* (apside) and, $\Gamma\hat{O}A = \varpi$. The mean planet P_0 moves along the deferent circle uniformly. Its longitude called *madhyamagraha* (mean longitude) is given by $\theta_M = \Gamma\hat{O}P_0$. Construct a circle of radius r with P_0 as the centre. This is the *mandanīcoccavṛtta* (epicycle) on which the *mandasphuṭa* is located. Draw a line from P_0 parallel to OA, which intersects the epicycle at P. This is the location of the *mandasphuṭa* whose longitude is given by $\theta_{MS} = \Gamma\hat{O}P$. This is the epicycle model.

Alternatively, locate O' on OA such that $OO' = r$. The circle with O as the centre and r as the radius is also known as *mandanīcoccavṛtta*. Draw a circle with radius R with O' as the centre. This is called the *pratimaṇḍala* or 'eccentric circle'. It is easy to see that the *mandasphuṭa*, P is located on this *pratimaṇḍala* such that $\Gamma\hat{O'}P = \Gamma\hat{O}P_0 = \theta_M$. That is, the planet moves uniformly around O', but not with respect to O at

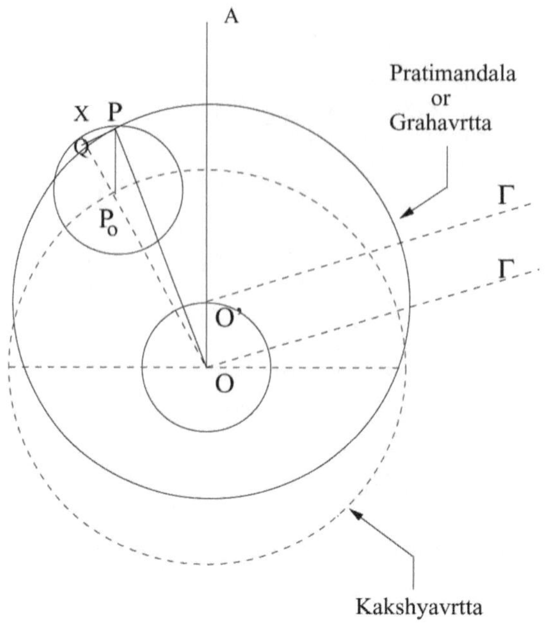

Fig. 2.2 *Manda-saṃskāra* or the correction due to eccentricity of the orbit

which the longitude is measured. This is the 'eccentric circle' model. Clearly, it is equivalent to the 'epicycle model'.

Draw PQ perpendicular to (extended) OP_0. Now, $PP_0 = r$. $P\hat{P_0}Q = \Gamma\hat{O}P_0 - \Gamma\hat{O}A = \theta_M - \varpi$ is the *manda-kendra*, or the anomaly. $P\hat{O}P_0 = \Gamma\hat{O}P_0 - \Gamma\hat{O}P = \theta_M - \theta_{MS}$. Now

$$PQ = r\sin(P\hat{P_0}Q) = OP\sin(P\hat{O}P_0). \tag{2.1}$$

$K = OP$ is called *mandakarṇa*. It is easily seen that

$$K = [(R + r\cos(\theta_M - \varpi))^2 + r^2\sin^2(\theta_M - \varpi)]^{1/2}. \tag{2.2}$$

In *Tantrasaṅgraha* (as also in *Āryabhaṭīya*, *Brāhmasphuṭasiddhānta*, etc.), r is not a constant but proportional K, such that

$$\frac{r}{K} = \frac{r_0}{R}, \tag{2.3}$$

where r_0 is prescribed in the text for each planet. It is not a constant but depends on the anomaly itself. Now K depends on r, and r is proportional to K. In *Tantrasaṅgraha* and *Yuktibhāṣā*, an iterative procedure is given to find K. K can be found non-iteratively also, using an elegant geometrical method. The exact expression for the *mandakarṇa*, K is:

$$K = \frac{R^2}{\sqrt{R^2 - r_0^2\sin^2(\theta_M - \varpi)} - r_0\cos(\theta_M - \varpi)}, \tag{2.4}$$

so it is determined in terms of known quantities. This exact expression is attributed to Mādhava by Nīlakaṇṭha. As r is not a constant, the planet's orbit is not a circle, but some kind of an oval. Taking Eq. 2.3 into account, we get

$$\sin(\theta_{MS} - \theta_M) = -\frac{r_0\sin(\theta_M - \varpi)}{R}. \tag{2.5}$$

So, one does not need to know the *mandakarṇa* to find the correction $\theta_{MS} - \theta_M$. As the mean longitude θ_M which varies linearly with time can be readily found at any instant, and the *mandocca*, ϖ is known, the *mandaphala*, $\theta_{MS} - \theta_M$ can be found from the above equation. Adding this to the mean longitude θ_M, we obtain the *mandasphuṭagraha*, θ_{MS}. This is the *sphuṭagraha* or the true longitude in the case of the sun, and essentially so for the moon. For these, O is the *Bhagola-madhya* or the centre of the earth.

The *mandasphuṭagraha* is the true heliocentric longitude for the *tārāgraha*s (actual planets), as we will see shortly. One more correction called *śīghra-saṃskāra* has to be applied to them to obtain their geocentric longitudes. When $\frac{r_0}{R} \ll 1$, $\sin(\theta_{MS} - \theta_M) \approx \theta_{MS} - \theta_M$ and Eq. 2.5 reduces to

$$\theta_{MS} - \theta_M = -\frac{r_0 \sin(\theta_M - \varpi)}{R}. \tag{2.6}$$

Comparing this with Eq. 2.35 in the Appendix, it can be seen that this has the same form as the equation of centre to $O(e)$ in Kepler's model, with $r_0/2R$ playing the role of eccentricity, e. This equivalence is only to the first order in e. The equation of centre has different forms in higher orders of e in Kepler's model and Indian models.

2.5.2 Conversion to the geocentric frame Śīghra-saṃskāra

Exterior planets

The following explicit geometrical construction is described in *Yuktibhāṣā* to obtain the geocentric longitude of an exterior planet. The final formula is essentially the same as in earlier texts.

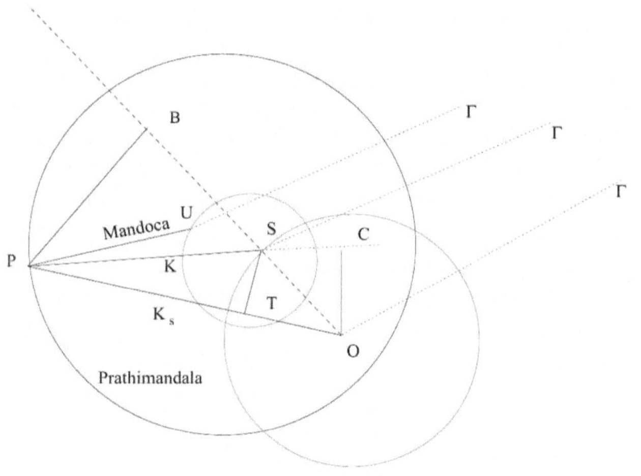

Fig. 2.3 *Śīghra-saṃskāra* for exterior planets to obtain the geocentric longitude

In Figure 2.3, O is the *Bhagola-madhya* or the centre of the earth. The *śīghra-nīcocca-vṛtta* or the *śīghra* epicycle is a circle with O as the centre, and whose radius, r_s is prescribed. The *śīghroccas*, S is located on this circle. It is also stated that *śīghroccas* is the *āditya-madhyama* (the mean sun). The *manda-nīcocca-vṛtta* or the *manda*-epicycle is a circle with the *śīghroccas* as the centre. The *mandocca*, U is located on this circle. The planet P is located on the *pratimaṇḍala* which is centered at U. $K = SP$ is the *mandakarṇa* and the circle with S as the centre and SP as the radius is the *manda-karṇa-vṛtta*. $P\hat{S}\Gamma$ is the *mandasphuṭa*, which is is the true heliocentric longitude. $P\hat{S}B$ which is the difference between the *mandasphuṭa* and *śīghroccas* is known as *śīghra-kendra*. $\Gamma\hat{O}P$ is the true geocentric planet known as the *śīghra-sphuṭa*. The *śīghra-sphuṭa* is found in the same manner from the *mandasphuṭa*, as the *mandasphuṭa* is found from the mean planet, *madhyama-graha*. Thus it may be noted that, in the computation of the *śīghra-sphuṭa*, the *śīghroccas* and the *manda-karṇa-vṛtta* will play the same roles as the *mandocca* and the *pratimaṇḍala* did in the computation of the *mandasphuṭa*. The *śīghra-karṇa* $K_s = OP$ can be determined in terms of the $SP = K$. Apart from the similarities, there is one difference. In *mandasamskāra*, we had noted that the radius of the epicycle r increases or decreases in the same way as K. In *śīghra-samskāra* , the radius $OS = r_s$, does not vary with the *śīghra-karṇa*, K_s. To start with, both the mean radius r_0 of the *manda*-epicycle and the radius r_s of the *śīghra*-epicycle are specified in the measure of the *pratimaṇḍala* radius, being *trijyā* or $R = 3438'$.

It would be useful to put these in terms of mathematical symbols for writing out the formula for the geocentric longitude.

$$\begin{aligned}
\textit{madhyama-graha} &= \Gamma\hat{U}P = \theta_M,\\
\textit{mandasphuṭa} &= \Gamma\hat{S}P = \theta_{MS},\\
\textit{mandakarṇa} &= SP = K,\\
\textit{śīghroccas}, \text{mean sun} &= \Gamma\hat{O}S = \theta_S,\\
\text{Radius of the } \textit{śīghra} \text{ epicycle} &= OS = r_s,\\
\textit{śīghra-kendra} &= P\hat{S}B = \Gamma\hat{S}P - \Gamma\hat{S}B \qquad (2.7)\\
\textit{śīghra-karṇa} &= OP = K_s,\\
\textit{śīghra-sphuṭa} &= \Gamma\hat{O}P = \theta_g. \qquad (2.8)
\end{aligned}$$

Here the *mandasphuṭa*, θ_{MS} and the *mandakarṇa*, K are determined in terms of the mean longitude, θ_M, *mandocca*, and the prescribed value of r_0, as described in the previous sub-section. The following formula for θ_g can be easily derived from the geometrical construction in Figure 2.3:

$$\begin{aligned}
\sin(\theta_g - \theta_{MS}) &= \frac{r_s \sin(\theta_{MS} - \theta_S)}{K_s},\\
&= \frac{r_s \sin(\theta_{MS} - \theta_S)}{[(K + r_s\cos(\theta_{MS} - \theta_S))^2 + r_s^2 \sin^2(\theta_{MS} - \theta_S)]^{1/2}}. \qquad (2.9)
\end{aligned}$$

This is the formula given in the earlier texts also, like *Sūryasiddhānta*, *Mahābhāskarīya*, etc. This formula of *Yuktibhāṣā* is exactly the same as would follow from the Keplerian model (see Eq. 2.41, of the Appendix) for an exterior planet, when θ_{MS} in the former is identified with the true heliocentric longitude in the latter, θ_h, and r_s/K in the former is identified with r/R in the latter. The identification of θ_{MS} with θ_h is clear enough, as θ_{MS} is the heliocentric mean planet to which the correction due to eccentricity is added. There is a small error as θ_s is the mean sun and not the true sun. This error is present in Ptolemy's *Almagest* (Toomer 1984), and even in *de Revolutionibus* of Copernicus (Sverdlow and Neugebauer 1984). We compare the average values of r_s/R which are the ratios of the earth-planet and sun-planet distances in the geometrical model with the ratios of the actual physical values of these in Table 2.2. There is reasonable agreement between the values.

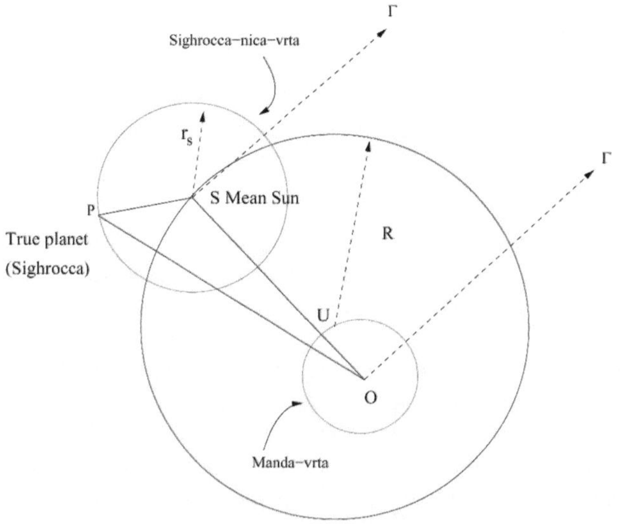

Fig. 2.4 *Śīghra-saṃskāra* for interior planets to obtain the geocentric longitude, prior to *Tantrasaṅgraha*

In both the Indian and Greek traditions, the mean sun was taken to be the mean planet for the interior planets (Mercury and Venus). This is due to the fact that the directions of Mercury and Venus are close to that of the sun, as viewed from the earth. So, their geocentric longitudes would not be very different from the longitude of the sun. The procedure for the computation of the geocentric longitude of an interior planet prior to *Tantrasaṅgraha* is represented geometrically in Figure 2.4. Here S is the mean sun moving in a circle around U, which is the *mandocca* of the planet with OU equal to the radius of the *manda*-epicycle of the planet. So, S is the mean sun, to which the *mandasamskāra* pertaining to the planet is applied. P is the *śīghroccas* which is actually the mean heliocentric planet, moving around S. Then the geocentric longitude of the inner planet is given by $\Gamma\hat{O}P$. This is obviously wrong. The *manda-sphuṭa* $\Gamma\hat{O}S$ here does not correspond to anything physical. The same mistake was committed by Ptolemy in his *Almagest* (Toomer 1984) and even by Copernicus (Sverdlow and Neugebauer 1984).

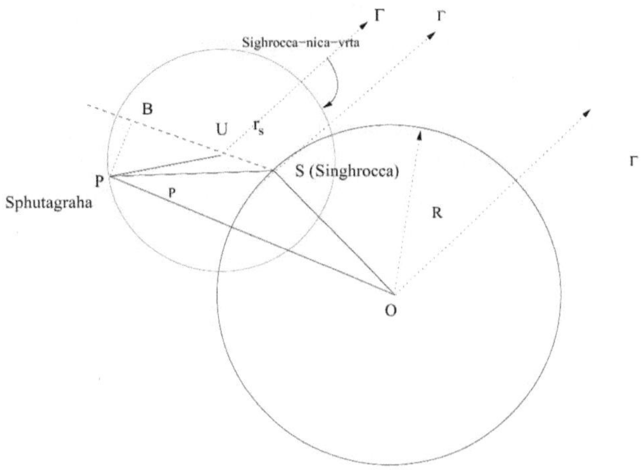

Fig. 2.5 *Śīghra-saṃskāra* for interior planets in the revised formulation of *Nīlakaṇṭha*

Nīlakaṇṭha Somayājī revised this picture in *Tantrasaṅgraha* and his other works (Ramasubramanian et al. 1994). His revised formulation is what is described in *Yuktibhāṣā* and depicted in Figure 2.5. Here the mean sun S is the *śīghroccas*, just as for exterior planets. S moves on a circle of radius R with the *Bhagola-madhya*, O as the centre. The planet P moves on an eccentric circle with the *mandocca*, U as the centre, with the prescribed value, r_s as the radius. SP is the *manda-karṇa* with its value reduced from K to $\tilde{r}_s = Kr_s/R$. $\Gamma\hat{S}P$ is the *manda-sphuṭa*, which is the longitude of P measured with respect to S. $\Gamma\hat{O}P$ is the *sīghra-sphuṭa*, which is the geocentric longitude of the planet. We define the symbols in the usual fashion:

$$\textit{madhyama-graha} = \Gamma\hat{U}P = \theta_M,$$
$$\textit{mandasphuṭa} = \Gamma\hat{S}P = \theta_{MS},$$
$$\textit{mandakarṇa} = SP = \tilde{r}_s = = r_s\frac{K}{R},$$
$$\textit{śīghroccas}, \text{mean sun} = \Gamma\hat{O}S = \theta_S,$$
$$\text{Radius of the } \textit{śīghra} \text{ epicycle} = UP = r_s,$$
$$\textit{śīghra-kendra} = P\hat{S}B = \Gamma\hat{S}P - \Gamma\hat{S}B \tag{2.10}$$
$$\textit{śīghra-karṇa} = OP = K_s,$$
$$\textit{śīghra-sphuṭa} = \Gamma\hat{O}P = \theta_g. \tag{2.11}$$

In the case of the exterior planet, the mean sun which is the *śīghroccas* moves in a smaller circle of radius r_s and the planet moves in a circle of radius equal to *trijyā*, R around the *mandocca*. Here also, the mean sun is the *śīghroccas*, which moves in the bigger circle of radius R, and the planet moves in a circle of smaller radius, r_s around the *mandocca*. The *mandasphuṭa*, θ_{MS} is calculated from θ_M in the usual fashion. The *śīghra-karṇa*, K_s is given by

$$K_s = [(R + \tilde{r}_s\cos(\theta_{MS} - \theta_S))^2 + (\tilde{r}_s)^2\sin^2(\theta_{MS} - \theta_S)]^{1/2}. \tag{2.12}$$

Then the geometry of Figure 2.5 implies the following formula for the true geocentric longitude, θ_g:

$$\sin(\theta_g - \theta_s) = \frac{\tilde{r}_s \sin(\theta_{MS} - \theta_S)}{K_s},$$
$$= \frac{\tilde{r}_s \sin(\theta_{MS} - \theta_S)}{[(R + \tilde{r}_s \cos(\theta_{MS} - \theta_S))^2 + \tilde{r}_s^2 \sin^2(\theta_{MS} - \theta_s)]^{1/2}}. \quad (2.13)$$

This is the same as Eq. 2.46 in the Appendix for the geocentric longitude of an interior planet in Kepler's model, if we identify θ_{MS} here, with the true heliocentric longitude, θ_h there, and $\tilde{r}_s/R$ in the above geometrical picture with r/R in Kepler's model. Again, it is clear from the figure that θ_{MS} is the true heliocentric longitude (if we ignore the difference between the mean sun and the true sun). Ignoring the eccentricity of the inner planet's orbit, we compare the average values of r_s/R which are the ratios of the sun-planet and earth-planet distances in the geometrical model, with the ratios of the actual physical values of these in Table 2.2. There is reasonable agreement, just as for the exterior planets.

Table 2.2 Comparison of $\frac{r_s}{R}$ in *Tantrasaṅgraha* for *śīghra-saṃskāra* with the modern values of the ratio of the mean values of the earth-sun and planet-sun distances for exterior planets and the inverse ratio for interior planets

Planet	*Tantrasaṅgraha* value	Modern value
Mercury	0.375	0.387
Venus	0.725	0.723
Mars	0.656	0.656
Jupiter	0.194	0.192
Saturn	0.106	0.105

2.5.3 Latitude of planets

Just as for longitudes, we have a unified theory of latitudes in *Yuktibhāṣā*. First, the heliocentric latitude is considered. In Figure 2.6, O is the earth and S is the sun moving around the earth in the plane of the ecliptic. P is the planet moving in a plane inclined to the ecliptic, and $K = SP$ which lies in this plane is the *mandakarṇa*. The *apakramavṛtta*, which is the ecliptic, and the orbit of the planet intersect at N, which is the ascending node.

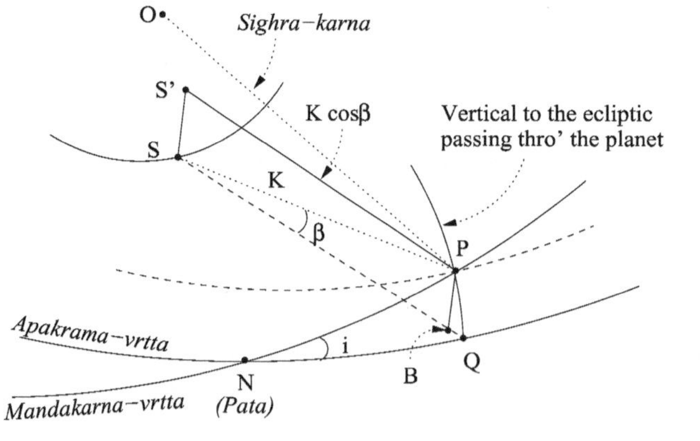

Fig. 2.6 The latitudinal deflection (*vikṣepa*) of a planet

The latitude of the planet is $\beta = PQ$ which is the arc perpendicular to the ecliptic passing through P. PB is perpendicular to the plane of the ecliptic. Then the *vikṣepa* of the planet is given to be

$$vikṣepa = v = PB = K \sin\beta. \quad (2.14)$$

Figure 2.7 described in *Yuktibhāṣā* relates the geocentric latitude $\tilde{\beta}$ with the heliocentric latitude β. We have

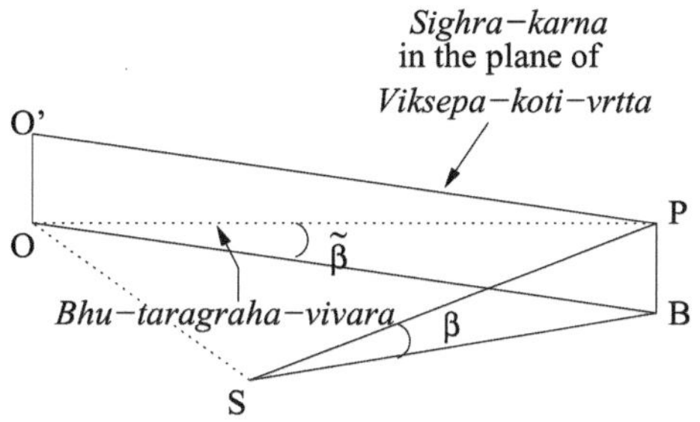

Fig. 2.7 Geocentric latitude, $\tilde{\beta}$

$$PB = OP\sin\tilde{\beta} = SP\sin\beta, \tag{2.15}$$

or,

$$\sin\tilde{\beta} = \frac{SP}{OP}\sin\beta. \tag{2.16}$$

When β and $\tilde{\beta}$ are small, this reduces to

$$\tilde{\beta} = \beta.\frac{SP}{OP}. \tag{2.17}$$

OP is the *bhū-tārā-graha-vivara* or the earth-planet distance, while SP is the sun-planet distance. This is exactly the same as the relation Eq. 2.51 in the appendix in Kepler's model, relating the geocentric latitude β_E and the heliocentric latitude β_s. Here it is pertinent to recall that the theory of latitude was totally unsatisfactory in *Almagest*, as the planes of the orbits were taken to be passing through the earth rather than the sun there (Toomer 1984). This was true of *de Revolutionibus* of Copernicus also (Sverdlow and Neugebauer 1984).

The latitudinal deflection affects the longitude also. The *vikṣepa-koṭi-vṛtta* is essentially *mandakarṇa-vṛtta* projected on to the plane parallel to the *śīghra-vṛtta* in which the planet is located. See Figures 2.6 and 2.7. It is pointed out that the geocentric longitude, *śīghra-sphuṭa* should be calculated taking the *vikṣepa-koṭi*, $K\cos\beta$ as the *mandakarṇa*. The result is the *graha-sphuṭa* (true planet) on the *śīghra-karṇa-vṛtta*, which is a circle with O' as the centre. O' is the point in the plane of *vikṣepa-koṭi-vṛtta*, at distance $OO' = PB$ from the ecliptic. This is the "reduction to the ecliptic" which was taken note of by some of the Islamic astronomers and later discussed by Tycho Brahe in late 16th century.

2.5.4 Summary of the Planetary Model in Yuktibhāṣā

1. The algorithms for the calculation of geocentric longitudes of planets, prior to *Tantrasaṅgraha* and *Yuktibhāṣā* more or less simulated the results of the Kepler model, but for the fact that the equation of centre for interior planets was wrong. But there are no clear explanations in the earlier texts.
2. Nīlakaṇṭha's revised formulation of the equation of centre for inner planets led him to an unified treatment of exterior and interior planets. The clearest articulation of this is to be found in *Yuktibhāṣā*. The planets move in eccentric orbits around *śīghroccas* (mean sun), which itself goes around the earth. The formulae for the geocentric longitudes and latitudes of planets here are essentially the same as in Kepler model.
3. Ptolemy's model for planetary motion is more complicated. Apart from the error in the formulation of equation of centre of inner planets, the theory for planetary latitudes is totally off the mark here. In fact, the same is true of the Copernican model also, revolutionary as it was in treating the earth also as a planet.
4. It should be mentioned that the geometrical model described in *Yuktibhāṣā* is for each individual planet separately. It does not give an unified picture involving all the planets, to scale.

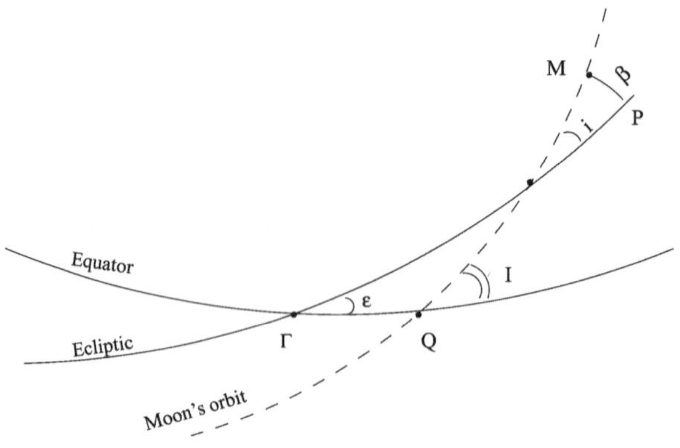

Fig. 2.8 Inclination of Moon's orbit

2.6 Inclination of moon's Orbit with Equator

Consider the orbit of the moon, which is inclined at an angle i with the ecliptic, which is itself inclined to the equator by an angle ε, in Figure 2.8. The declination of the moon is given by

$$\sin\delta_m = \cos\varepsilon\ \sin\beta + \sin\varepsilon\ \cos\beta\ \sin\lambda, \tag{2.18}$$

where $\lambda = \Gamma P$ and β are the longitude and latitude of the moon, respectively. In *Tantrasaṅgraha*, an alternate expression for δ_m in terms of the instantaneous inclination, I of the moon's orbit is given. We describe the derivation of the expression for I as presented in *Yuktibhāṣā*.

moon's orbit is called the *vikṣepavṛtta* and the northern pole of this denoted by V is called *vikṣepa-pārśva*. The *vikṣepa-vṛtta* intersects the ecliptic at *Rāhu* (ascending node of the moon) and *Ketu* (descending node) and diverge northwards and southwards respectively from these points. Consider the situation when the *Rāhu* coincides with the vernal equinox Γ.

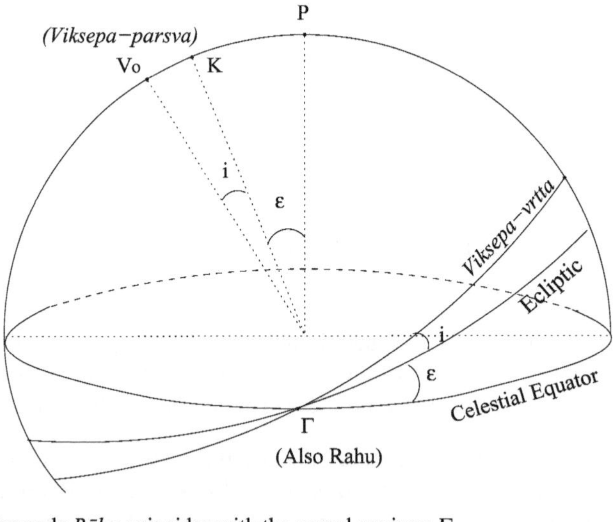

Fig. 2.9 Moon's orbit when the node *Rāhu* coincides with the vernal equinox Γ

This is depicted in Figure 2.9. Here P, K and V_0 are the poles of the equator, ecliptic and the *vikṣepa-vṛtta*, respectively. In this situation, it is clear that the inclination of moon's orbit with the equator is $I = \varepsilon + i$.

Consider the situation when the longitude of *Rāhu* is λ_N. Let the northern pole of moon's orbit be at V. The relative locations of V_0, V, K and P are shown in Figure 2.10(b). OP, OK and OV are perpendiculars to the planes of the equator, ecliptic and moon's orbit, respectively. The angle between OV and OP is the

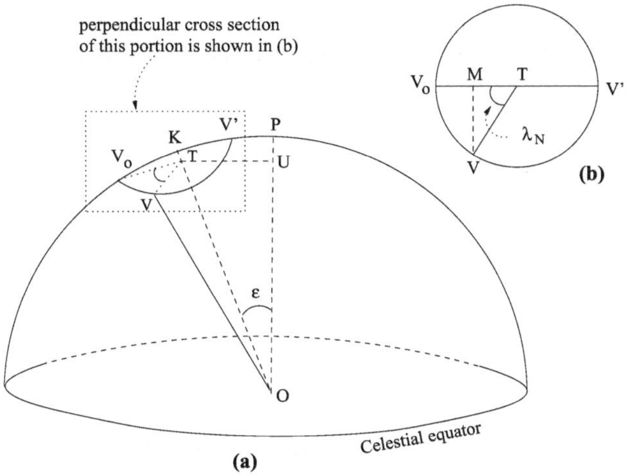

Fig. 2.10 The relative locations of the poles of the equator, ecliptic and moon's orbit

inclination of moon's orbit with the equator, I. V_0 is separated from K by the angle i. Drop a perpendicular V_0T from V_0 to OK, where O is the centre of the sphere. As the arc $V_0K = i$, $V_0T = R\sin i$. Draw a circle with radius $R\sin i$ centred at T in the plane perpendicular to OT. This circle called *vikṣepa-pārśva-vṛtta* (shown separately in Figure 2.10(b)) would be parallel to the plane of the ecliptic. Mark a point on this circle such that the angle corresponding to the arc V_0V is the longitude of *Rāhu*, λ_N. Drop a perpendicular TU from T to the '*akṣa-daṇḍa*' OP. Draw VM perpendicular to V_0T. Now $V\hat{O}T = i$. Hence,

$$\begin{aligned} OT &= R\cos i, \\ VT &= R\sin i, \\ TU &= OT\sin\varepsilon = R\cos i\ \sin\varepsilon, \\ VM &= VT\sin\lambda_N = R\sin i\ \sin\lambda_N, \\ \text{and } MT &= VT\cos\lambda_N = R\sin i\ \cos\lambda_N. \end{aligned} \tag{2.19}$$

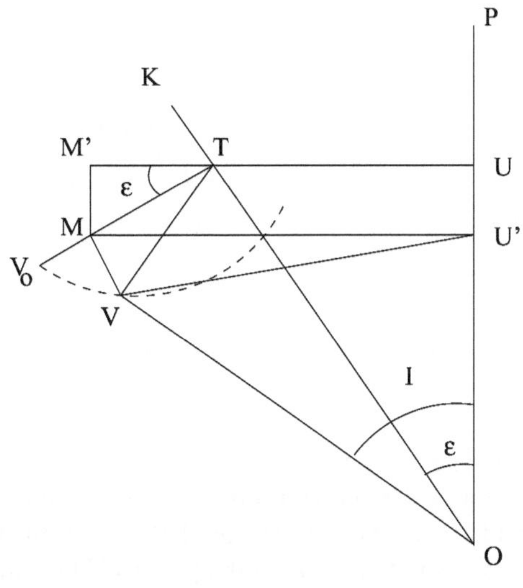

Fig. 2.11 The inclination of moon's orbit with the equator

In Figure 2.11, the points V_0, V, T (centre of the *vikṣepa-pārśva-vṛtta*), M and U have the same significance as in Figure 2.10. MV is perpendicular to the plane of the figure. Draw MU', perpendicular to the *akṣa-daṇḍa*, OP. VM is perpendicular to the plane of the figure and hence to OP, and MU' is also perpendicular to OP. Hence $VU'M$ is a triangle, right angled at M, and in a plane perpendicular to OP. Therefore, VU' is perpendicular to OP and is the desired distance, $R\sin I$, between V and *akṣa-daṇḍa*. Let MM' be perpendicular to UM' which is the extension of UT. The angle between TM' and TM is ε. It is clear that $MU' = M'U$. Therefore,

$$\begin{aligned} M'U &= M'T + TU \\ &= MT\cos\varepsilon + R\cos i\ \sin\varepsilon \\ &= R\sin i\ \cos\lambda_N\ \cos\varepsilon + R\cos i\ \sin\varepsilon. \end{aligned} \tag{2.20}$$

Then,

$$\begin{aligned} VU' &= \sqrt{(MV)^2 + (MU')^2} \\ &= \sqrt{(R\sin i\ \sin\lambda_N)^2 + (R\sin i\ \cos\lambda_N\ \cos\varepsilon + R\cos i\ \sin\varepsilon)^2}. \end{aligned} \tag{2.21}$$

Clearly $VU' = R\sin I$, where I, which is the angle corresponding to the arc VP, is the inclination of moon's orbit with the equator. Hence,

$$R\sin I = \sqrt{(R\sin i\ \sin\lambda_N)^2 + (R\sin i\ \cos\lambda_N\ \cos\varepsilon + R\cos i\ \sin\varepsilon)^2}. \tag{2.22}$$

2.7 Distance between the Centres of the Solar and Lunar Discs

As another example of the geometrical methods in *Yuktibhāṣā*, we discuss the distance between the centres of the solar and lunar discs. Now, by definition, the sun moves on the ecliptic when viewed from the centre of the earth. However, for an observer on the surface of the earth, the sun is displaced from the ecliptic, because of parallax. The arc corresponding to this displacement is called '*nati*'. The moon is deflected from the ecliptic, both due to its latitude and parallax. The net deflection due to these factors is termed *vikṣepa*.

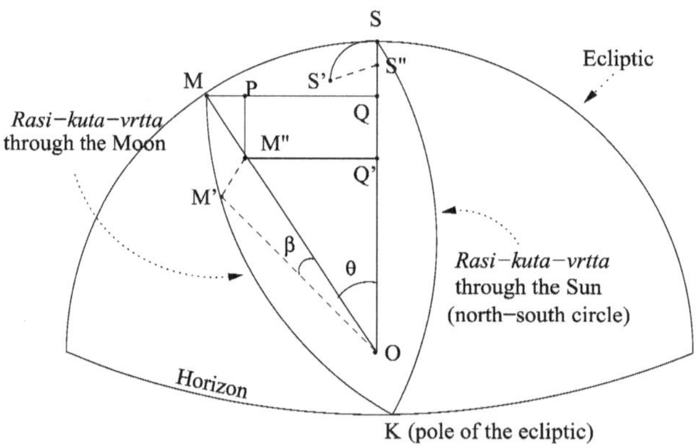

Fig. 2.12 Calculation of *bimbāntara*, the distance between the solar and lunar discs

In Figure 2.12, O is the observer. The sun without *nati* is conceived to be at the zenith and the ecliptic is conceived as the prime vertical with its poles on the north and south points. The moon without *vikṣepa* is at M. The *rāśi-kūṭa-vṛtta*, or the secondary to the ecliptic through the sun is the north-south circle. (In the figure, the ecliptic which is the prime vertical is in the plane of the paper.) If O is the centre of the sphere, OS is the vertical line. R is the radius of the sphere.

The sun with *nati* is at S', and the moon with *vikṣepa* is at M'. $S'S''$ and $M'M''$ are drawn perpendicular to OS and OM, respectively. MQ and $M''Q'$ drawn perpendicular to OS, and $M''P$ to MQ. We define:

$$\text{Difference in longitudes of sun and moon} = M\hat{O}S = \theta,$$
$$\text{Sun's } nati = \text{Arc } SS' = \eta_s,$$

$$\text{Arc of moon's } vik\d{s}epa = M\hat{O}M' = \beta. \tag{2.23}$$

Now the vertical distance between the true moon (M') and the true sun (S') is termed the first *rāśi*, r_1. It is straightforward to show that

$$\begin{aligned} r_1 = S''Q' &= S''Q + QQ' \\ &= R(1-\cos\theta) - R(1-\cos\eta_s) + R\cos\theta(1-\cos\beta). \end{aligned} \tag{2.24}$$

The horizontal distance $M''Q'$ between the sun and the moon in the plane of the ecliptic is termed the second *rāśi*, r_2. It is given by

$$r_2 = M''Q' = R(1-\cos\theta\,\cos\beta). \tag{2.25}$$

The difference between the sun and the moon along the line perpendicular to the ecliptic is termed the third *rāśi*, r_3. It is given by

$$r_3 = M'M'' - S'S'' = R\sin\beta - R\sin\eta_s. \tag{2.26}$$

Then the *bimbāntara*, or the distance $S'M'$ between the centres of the lunar and solar discs is given by

$$bimb\bar{a}ntara = d = \sqrt{r_1^2 + r_2^2 + r_3^2}. \tag{2.27}$$

This is how the distance would be calculated in modern three dimensional coordinate geometry also, as r_1, r_2 and r_3 are the differences in the coordinates of the centres of the lunar and solar discs along three mutually perpendicular directions.

2.8 Concluding Remarks

Yuktibhāṣā is written in the style of a text book. Indeed its mathematics part has been hailed as the first text book of calculus. Its astronomy part whose aim is to elucidate the contents of *Tantrasaṅgraha* is also written in the style of a text book, perhaps for the first time for a work on spherical astronomy in India. *Tantrasaṅgraha* is a comprehensive compendium of the algorithms of Indian astronomy, with several improvisations in presentation, revision of the planetary model, more systematic treatment of several problems, especially in spherical trigonometry, and replacement of many earlier approximate results in the Indian tradition with exact ones. *Yuktibhāṣā* explains all these algorithms. It does more than that. It lays down the theoretical basis for each topic, before setting out the explanations of the results. No planetary parameters are mentioned in the text, nor are there illustrative examples of the computational procedures. The aim of the author is to equip the student with the theoretical basis and the explanations of the algorithms, rather than train him in the use of them.

It is intriguing that there is no other comparable original work in any regional language in India, before or after *Yuktibhāṣā*, at least in the field of mathematics and astronomy. It has been pointed out that its mathematics part is very effective in the use of an almost totally conversational Malayalam in conveying the subtle and complex reasoning that this new knowledge is based on Divakaran 2007. This is true of the astronomy part also. We have only attempted to give an outline of this part, with detailed discussions of some topics in the work, so that the reader can get a feel of the approach to astronomical problems in the 16th century India in general, and Kerala in particular.

Acknowledgements Much of this chapter is based on the explanatory notes of *Yuktibhāṣā*, which was the combined effort of K. Ramasubramanian (KR), M. D. Srinivas (MDS) and myself. I thank KR and MDS for a wonderful collaboration, and for offering several suggestions for improving the present chapter. I also thank late K.V. Sarma who edited and translated *Yuktibhāṣā*, and who was a source of inspiration.

References

1. Divakaran PP (2007) The first text book of calculus: *Yuktibhāṣā*. Journal of Indian Philosophy. 35:417
2. Pillai SK (1958) *Tantrasaṅgraha of Nīlakaṇṭha with the Prose Commentary Laghuvivṛtti by* Śaṅkara Vāriyar, Ed., Trivandrum Sanskrit Series no.188 Trivandrum Sarma KV(1977) *Tantrasaṅgraha with the commentary Yuktidīpikā by* Śaṅkara Vāriyar, Ed., V.V.B. Institute, Panjab University, Hoshiarpur
3. Ramasubramanian K. and Sriram MS. *Tantrasaṅgraha of Nīlakaṇṭha Somayājī*, Translation with explanatory notes, in press
4. Ramasubramanian K, Srinivas MD, and Sriram MS (1994) 'Modification of the earlier Indian planetary theory by the Kerala astronomers (c.1500 AD) and the implied heliocentric picture of planetary motion', Current Science, **66**, 784; for details see Vol.II of *Ga-yukti-bhāṣā*
5. Ramavarma (Maru) Tampuran HH,and Akhileswarayyar AR (1948) *Yuktibhāṣā*, Eds, Mangalodayam Ltd. Thrissur
6. Sarma KV (1972) *A History of the Kerala School of Hindu Astronomy (in perspective)*, Vishveshvaranand Institude, Hoshiarpur
7. Sarma KV, and Hariharan S (1991)'*Yuktibhāṣā* of Jyeṣṭhadeva', Indian Journal of History of Science, **26**, 2
8. Sarma KV, and Narasimhan VS (1998) *Tantrasaṅgraha of Nīlakaṇṭha Somayājī*, Text (KVS) and Translation (VSN), Supplement to the Indian Journal of History of Science, INSA, New Delhi, **33** pp.148
9. Sarma KV (2004) *Gaṇita-Yukti-Bhāṣā of Jyeṣṭhadeva*, Ed. Vol.III. Sanskrit text, Indian Institute of Advanced Study, Shimla
10. Sarma KV, Ramasubramanian K, Srinivas MD, and Sriram MS (2009) *Gaṇita-Yukti-Bhāṣā of Jyeṣṭhadeva*, Edition of Malayalam text and English translation (KVS) and Explanatory notes (KR, MDS, MSS), Vol.I: Mathematics, Vol.II: Astronomy, Hindustan Book Agency, New Delhi, 2008; Rep. Springer Verlag, new York
11. Sen SN. 'Astronomy' in D. M .Bose, S. N. Sen and BV Su-bbarayappa (1986) *A Concise History of Science in India*, Indian National Science Academy, New Delhi, 1971; S.N.Sen and K.S.Shukla, Eds. *History of astronomy in India*, Indian National Science Academy, New Delhi and references therein; for a popular account see S.Balacha-ndra Rao (2008) *Indian Astronomy - an Introduction*, Universities Press, Hyderabad, (2000) B.V.Subbarayappa, *The Tradition of Astronomy in India, Jyotiḥśāstra*, Vol.IV, Part 4, Centre for Studies in Civilizations, New Delhi
12. Shukla KS, and Sarma KV (1976)*Āryabhaṭīya*, Ed. with translation and notes, Indian National Science Academy, New Delhi
13. Srinivas MD (1990), 'Indian approach to Science: The case of *Jyotiḥśāstra'*, Nos. 19-20, PPST Bulletin, Jume Chennai
14. Sriram MS (2002) 'Planetary models in Indian and Greek astronomical traditions' in *500 years of Tantrasaṅgraha: A Landmark in the History of Astronomy*, Eds. M.S.Sriram, K.Ramasubramanian and M.D.Srinivas,Indian Institute of Advanced Study, Shimla
15. Sverdlow NM, and Neugebauer O (1984) *Mathematical Astronomy in Copernicus's De Revolutionibus*, Springer Verlag, New York
16. Toomer GJ (1984) *Ptolemy's Almagest*, Duckworth, London
17. Whish CM (1834) 'On the Hindu quadrature of the circle, and the infinite series of the proportion of the circumference of the diameter exhibited in the *four shastras*, the *Tantrasangraham, Yucti Bhasa, Carana Paddhati and Sadratnamala*', Transactions of the Royal Asiatic Society of Great Britain and Ireland, **3**, 509

Appendix

Geocentric Position of a Planet in Kepler's Model

The planetay models in ancient times can be appreciated better, if we understand how the geocentric position of a planet is calculated in Kepler's model, which is correct in its essentials even today (Sriram et al. 2002). The three laws discovered by Kepler in the early 17th century form the basis of our present understanding of planetary orbits. The three laws are:
1. Each planet moves around the sun in an ellipse, with sun as one of the foci.
2. The areal velocity of a planet in its orbit is a constant.
3. The square of the time period of a planet in its orbit is proportional to the cube of the semi-major axis of the ellipse in which it moves.

Kepler's laws can be derived from Newton's second law of motion and law of gravitation[3]. If we take the time period of a planet as an input, Kepler's third law does not play any role for an individual orbit. We should note that Kepler's laws are essentially kinematical laws, which do not make any reference to the concepts of 'acceleration' and 'force', as we understand them. Even then, they capture the very essence of the nature of

[3] Conversely, Newton's inverse square law of gravitation can also be arrived at from Kepler's laws!

planetary orbits and can be used to calculate the planetary positions, once we know the parameters of the ellipse and the initial positions. The planetary models in ancient astronomy are kinematical and should be really compared with Kepler's model. In the following, we elaborate on the computation of the geocentric longitude and latitude of a planet in Kepler's model.

Elliptic Orbits and Equation of Centre

The elliptic orbit of a planet (P) around the sun (S) is represented in Figure 2.13 a and b are the semi-major and semi-minor axes of the ellipse. Γ refers to the first point of Aries. $\varpi = \Gamma\hat{S}A$ is the longitude of the aphelion (A). $\theta_h = \Gamma\hat{S}P$ is the heliocentric longitude of the planet.

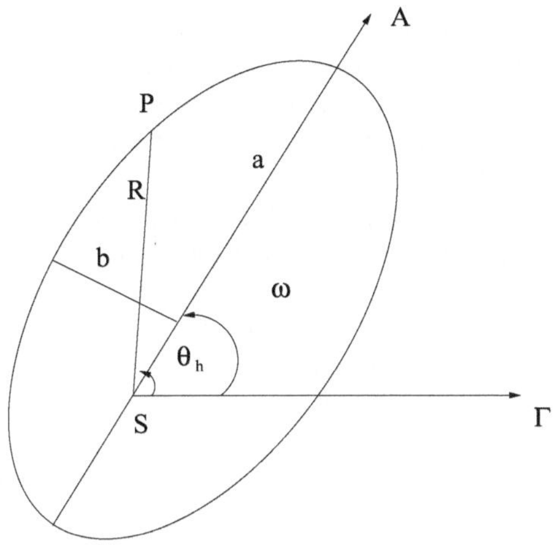

Fig. 2.13 Elliptic orbit of a planet around the sun

Here e is the eccentricity of the ellipse. a and b are the semi-major and semi-minor axes, and $l^2 = a^2(1-e^2)$. Then the equation of the ellipse is written as

$$\frac{l}{r} = 1 - e\cos(\theta_h - \varpi). \tag{2.28}$$

Therefore,

$$r = l[1 + e\cos(\theta_h - \varpi)] + O(e^2), \tag{2.29}$$

$$r^2 = l^2[1 + 2e\cos(\theta_h - \varpi)] + O(e^2). \tag{2.30}$$

Now the areal velocity of the planet at any instant is $\frac{1}{2}r^2\dot{\theta}_h$. According to Kepler's second law, it is a constant. As the area of an ellipse is πab, the areal velocity can also be written as $\frac{\pi ab}{T} = \frac{\omega ab}{2}$, where T is the time period and $\omega = \frac{2\pi}{T}$ is the mean angular velocity of the planet. Hence,

$$r^2\dot{\theta}_h = \omega ab. \tag{2.31}$$

Using the expression for r^2 in Eq 2.30, we find

$$l^2\dot{\theta}_h[1 + 2e\cos(\theta_h - \varpi)] = \omega ab + O(e^2). \tag{2.32}$$

Now $l^2 = a^2\ (1-e^2) = a^2 + O(e^2)$ and $ab = a^2 + O(e^2)$. Hence,

$$\dot{\theta}_h[1 + 2e\cos(\theta_h - \varpi)] \approx \omega, \tag{2.33}$$

where the equation is correct to $O(e)$. Integrating with respect to time, we find

$$\theta_h + 2e\sin(\theta_h - \varpi)] \approx \omega t, \tag{2.34}$$

or again to $O(e)$,

$$\begin{aligned}\theta_h - \theta_M = \theta_h - \omega t &\approx -2e\sin(\omega t - \varpi), \\ &= -2e\sin(\theta_M - \varpi).\end{aligned} \tag{2.35}$$

Here $\theta_M = \omega t$ is the mean longitude which increases linearly with time, t. $\omega t - \varpi$, that is the difference between the longitudes of the mean planet and the apogee is the 'anomaly'. Eq. 2.35 gives the equation of centre which is the difference between the true heliocentric longitude θ_h and the mean longitude θ_M, to $O(e)$, in terms of the anomaly. Clearly, the equation of centre is a consequence of the eccentricity of the orbit.

Geocentric Longitude of an Exterior Planet

The orbits of all the planets are inclined at small angles to the plane of earth's orbit around the sun, known as ecliptic. We will ignore these inclinations and assume that all the planetary orbits lie in the ecliptic plane for the calculation of planetary longitudes, as the corrections introduced by them (inclinations) are known to be small. We will consider the longitude of an exterior planet like Mars, Jupiter or Saturn and and interior planet like Mercury or Venus, separately.

The elliptic orbit of an exterior planet (P) and that of the Earth (E) around the sun (S) are shown in Figure 2.14. Here, $\theta_h = \Gamma\hat{S}P$ is the true heliocentric longitude of the planet. $\theta_S = \Gamma\hat{E}S$ and $\theta_g = \Gamma\hat{E}P$ are the true geocentric longitudes of the sun and the planet, respectively, while r and R are the distances of the earth and the planet from the sun, which vary along their orbits. To facilitate comparison with the Indian models, $EP' = R$ is drawn parallel to SP. Then $P'P$ is parallel to ES and $P'P = r$. We have already described how θ_h is computed using the expression for the equation of centre. From this, the true geocentric longitude, θ_g has to be computed. Now

$$E\hat{P}S = P\hat{E}P' = \theta_g - \theta_h, \tag{2.36}$$

and

$$E\hat{S}P = 180^\circ - (\theta_S - \theta_h). \tag{2.37}$$

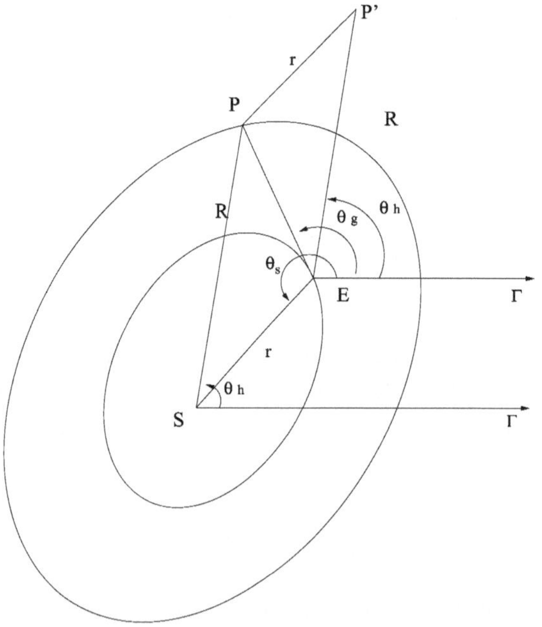

Fig. 2.14 Heliocentric and geocentric longitudes of an exterior planet in Kepler's model

In the triangle ESP,

$$EP^2 = R^2 + r^2 - 2rR\cos[180^\circ - (\theta_S - \theta_h)], \tag{2.38}$$

or,

$$EP = [(R + r\cos(\theta_S - \theta_h))^2 + r^2 \sin^2(\theta_S - \theta_h)]^{1/2} \tag{2.39}$$

Also,

$$\frac{\sin(E\hat{P}S)}{ES} = \frac{\sin(E\hat{S}P)}{EP}. \tag{2.40}$$

Using Eq. 2.36–Eq. 2.40,

$$\sin(\theta_g - \theta_h) = \frac{r\sin(\theta_S - \theta_h)}{[(R + r\cos(\theta_S - \theta_h))^2 + r^2 \sin^2(\theta_S - \theta_h)]^{1/2}}. \tag{2.41}$$

Here $(\theta_S - \theta_h)$, the difference between the longitudes of the sun and the heliocentric planet is the 'solar anomaly'. Thus, Eq. 2.41 gives $\theta_g - \theta_h$ in terms of the solar anomaly. Adding this to θ_h, we get the true geocentric longitude, θ_g of the planet.

Geocentric Longitude of an Interior Planet

The elliptic orbit of an interior planet (P) and that of the earth (E) around the sun are shown in Figure 2.15. Here, $\theta_h = \Gamma\hat{S}P$ is the true heliocentric longitude of the planet, which can be computed from the mean heliocentric longitude and the equation of centre. $\theta_S = \Gamma\hat{E}S$ and $\theta_g = \Gamma\hat{E}P$ are the true geocentric longitudes of the sun and the planet, respectively. Here r and R are the variable distances of the planet and the earth from the sun, respectively. Note the change in nomenclature, compared with the one for an exterior planet. Now

$$S\hat{E}P = \theta_g - \theta_S. \tag{2.42}$$

It can be easily seen that

$$E\hat{S}P = 180^\circ - (\theta_h - \theta_S), \tag{2.43}$$

$$EP = [(R + r\cos(\theta_h - \theta_S))^2 + r^2 \sin^2(\theta_h - \theta_S)]^{1/2}. \tag{2.44}$$

Also,

$$\frac{\sin(S\hat{E}P)}{SP} = \frac{\sin(E\hat{S}P)}{EP}. \tag{2.45}$$

Using Eq. 2.42–Eq. 2.45,

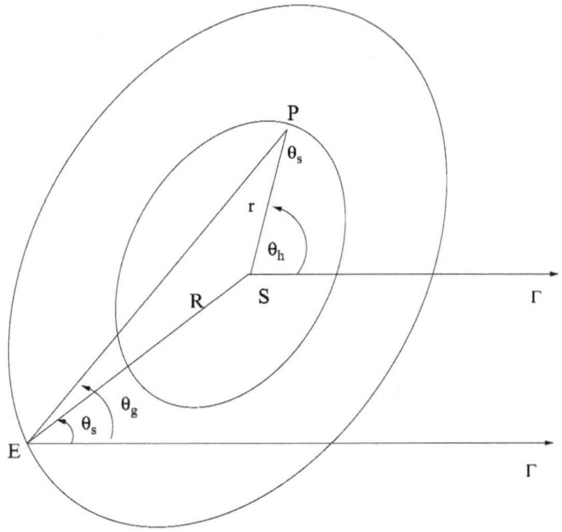

Fig. 2.15 Heliocentric and geocentric longitudes of an interior planet in Kepler's model

$$\sin(\theta_g - \theta_S) = \frac{r\sin(\theta_h - \theta_S)}{[(R + r\cos(\theta_h - \theta_S))^2 + r^2\sin^2(\theta_h - \theta_S)]^{1/2}}. \tag{2.46}$$

The difference $(\theta_g - \theta_S)$ is determined from this equation. Adding this to θ_S, we get the true geocentric longitude, θ_g of the planet. Note that the true longitude of the sun, θ_S, and the true heliocentric longitude of the planet, θ_h (obtained by adding the equation of centre for the planet to the mean heliocentric longitude longitude) should be obtained first. Then the true geocentric longitude of an interior planet can be obtained using Eq. 2.46.

Heliocentric and Geocentric Latitudes of a Planet

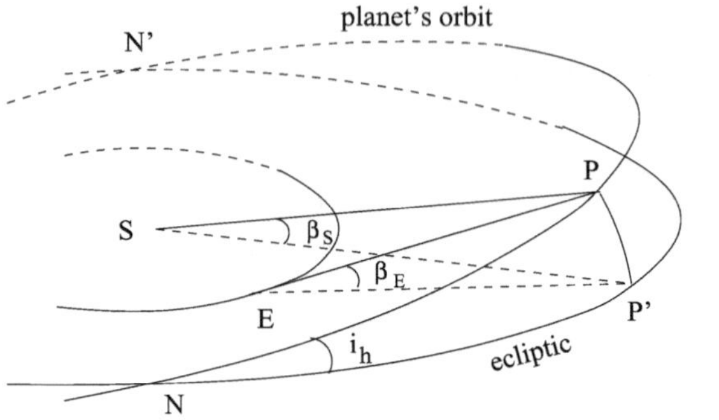

Fig. 2.16 Heliocentric and geocentric latitudes of a planet in Kepler's model

In Figure 2.16, the orbit of the planet (P) is inclined at an angle i_h to the ecliptic. N and N' are the nodes. PP' is the circular arc perpendiular to the ecliptic. Then the heliocentric latitude β_S is given by

$$\beta_S = PSP' = \frac{PP'}{SP}. \tag{2.47}$$

If λ_P and λ_N are the heliocentric longitudes of the planet and the node, it can be shown that

$$\sin\beta_S = \sin i_h \sin(\lambda_P - \lambda_N), \tag{2.48}$$

or,

$$\beta_S \approx i_h \sin(\lambda_P - \lambda_N), \tag{2.49}$$

as i_h and β_S are small. Note that λ_P stands for the true heliocentric longitude, whether it is an exterior or an interior planet. We have also shown earth's orbit in Figure 2.16. The geocentric latitude β_E is given by

$$\beta_E = PEP' = \frac{PP'}{EP}. \tag{2.50}$$

From Eq. 2.47 to Eq. 2.50, we find that

$$\begin{aligned}\beta_E &= \beta_S\frac{SP}{EP} \\ &= \frac{i_h\, SP\, \sin(\lambda_P - \lambda_N)}{EP},\end{aligned} \tag{2.51}$$

where EP, the true distance of the planet from the earth, can be found from Eq. 2.39 or Eq. 2.44.

Chapter 3
General Relativity and Astrophysics

Jayant V. Narlikar

3.1 Introduction

In 1915 Einstein published the final version of his basic general theory of relativity. Although he was to further modify it in 1917 by adding the so-called 'lambda term' the framework of the theory for local applications remained the same. For our reference we will take the form of the 1915 field equations as

$$R_{ik} - \frac{1}{2}g_{ik}R = -\frac{8\pi G}{c^4}T_{ik}, \tag{3.1}$$

while those with the lambda term are:

$$R_{ik} - \frac{1}{2}g_{ik}R + \lambda\, g_{ik} = -\frac{8\pi G}{c^4}T_{ik}, \tag{3.2}$$

Here the spacetime coordinates are x^k, $k = 0, 1, 2, 3$, with 0 usually standing for the timelike dimension. The spacetime metric is g_{ik}. The signature of the metric is [+,–,–,–]. T_{ik} denotes the energy momentum tensor of the physical entities present in spacetime.

As is well-known, the general theory of relativity was considered esoteric by most physicists and for a considerable period of time very few people worked on it. The 1919 eclipse results on the bending of light, announced by Eddington went a long way towards gaining crdibility for the theory. The eclipse success also excited the popular mind while the media had a field day broadcasting news items related to the event.

In this connection it is worth recalling that in 1919, shortly after the announcement of the results by Eddington a popular account of the whole experiment was published by the *Statesman*, Calcutta on November 13, 1919 and the account was written by Meghnad Saha, then a young physics lecturer entering into the newly emerging field of astrophysics (AP). Indeed it would not be an exaggeration to say that Saha's ionization equation provided the basic foundations on which the theory of stellar atmospheres is based and which in turn leads to the theory of stellar structure, considered as central to AP.

3.2 The Calcutta School

However, general relativity (GR) came to India through the works of two applied mathematicians, Nikhil Ranjan Sen at Calcutta University and Vishnu Vasudeva Narlikar at the Banaras Hindu University. Sen was a contemporary of two other distinguished scientists of the future: Meghnad Saha and Satyendra Nath Bose at the Presidency College, Kolkata. In 1913 all three passed their B.Sc. examination with honours in mathematics, Bose standing first, Saha second and Sen third in order of merit.

For his academic career and research Sen opted for applied mathematics and theoretical physics. In 1921 he was awarded his first doctorate (D.Sc.) by Calcutta University, while holding a teaching position at the University's applied mathematics department. Subsequently he took study leave to visit and work at some of the well-known centres in Europe. His researches under von Laue on GR and cosmogony brought him a

Inter-University Centre for Astronomy and Astrophysics, Ganeshkhind, Pune–411 007. e-mail: jvn@iucaa.ernet.in second doctorate (Ph.D.), from Berlin University. In Europe Sen had the benefit of coming in contact with the leading physicists of the day including Max Planck, Albert Einstein, Arnold Sommerfeld and Louis de Broglie.

Returning to India in 1924, Sen was appointed to the Rashbehari Ghosh Chair of Calcutta University and set up a flourishing school in applied mathematics and GR. Several young researchers joined his school and emerged well-qualified applied mathematicians and relativists. Sen's own contributions may be highlighted as follows:

His solution of Einstein's equations for the gravitational field of a spherical shell is well known (Sen 1924). It is shown that by a coordinate transformation the metric of the interior of the shell is the same as that of the de Sitter universe (Sen and von Laue 1924).

Sen analysed the equilibrium of a charged particle using the field equations that Einstein had modified in 1919. Assuming a charge distribution inside a finite-sized particle, Sen showed that under spherical symmetry, three-fourth of the total energy of the particle is electrical and one-fourth gravitational. Thus the mass of the electron was seen to consist of two components: the Lorentz mass and the other based on the charge distribution not uniquely determined by the theory (Sen 1927).

Sen worked on cosmology also and had a model of the universe which was static but not perfectly spherical (Sen 1934, 1935a, 1935b). He showed that the total mass of the model (which had a finite extent) was higher than that of the spherically symmetric Einstein universe. A stability analysis using this property then showed that the static Einstein model would lead to the expanding universe.

Studies of equilibrium configurations of constant density spheres led Sen (1933, 1937) to the conclusion that in contrast to classical Newtonian spheres, the relativistic ones have a maximum radius and a minimum mass for a given constant density. For a radius less than the maximum radius there are two possible configurations. The central pressures in these configurations sandwich the pressure present in the configuration of maximum radius. For a density corresponding to the maximum density of matter, the largest equilibrium configuration has a radius of 133 km, whereas the sphere of maximum mass is some 24 times the mass of the sun (Sen 1941). These conclusions are of interest in the modern context of limits on stellar masses at the end of nuclear energy generation.

Sen (1954) with T.C. Roy also studied the gravitational field of a stationary star cluster in which, following Einstein, each star follows a circular trajectory. They found a singularity-free solution which could be fitted to an extenal expanding spacetime. The gravitational mass of the cluster decreases as the cluster contracts, it being equal to the rest mass at infinite separation. The minimum gravitational mass is utmost 5% less than the rest mass. The velocities of the outermost stars approach the speed of light as the minimum is approached. If one puts the constraints that astronomically observed velocities only are found, then the difference between the two masses is insignificant and one may use Newtonian dynamics and gravitation. This type of work would have been found very interesting had it received suitable exposure in the early 1960s when equilibrium of supermassive objects was under consideration.

Since this chapter is about GR and its astrophysical applications we will not describe Sen's other works in quantum mechanics, fluid dynamics, etc. His reputation as a good teacher attracted students who worked with him sharing all his research interests. Amongst these a little known relativist was B. Datt.

Datt's work on gravitational collapse is described in his paper written in German and published in a German journal (Datt 1938). The work anticipated by an year, the much cited work of Oppenheimer and Snyder (1939) on the same topic. Unfortunately very little is known about Datt and his subsequent work. This paper was submitted from the Presidency College, Calcutta and the author thanks Dr J. Ghosh for consultation. We briefly describe this work as it is important in the context of relativistic AP.

For a spherically symmetric line element to decribe a spherically symmetric mass expanding or contracting with time, Datt writes

$$ds^2 = e^{\nu} dt^2 - e^{\lambda} dr^2 - e^{\mu} r^2 (d\theta^2 + \sin^2\theta d\phi^2). \tag{3.3}$$

Here the coordinates (r, θ, ϕ) are comoving coordinates of a typical pointlike component of the spherical object. The time coordinate t likewise denotes the proper time of this point. Datt then sets up the field equations and solves them in as general a form as possible. He then deduces special cases like the collapse of a dust ball (having pressure-free matter) which is similar to the Friedmann-Lemaitre models. Landau and Lifshitz in their book the *Classical Theory of Fields* (Third Edition) describe Datt's solutions in some detail.

We will return to the problem of gravitational collapse and other related issues later in this chapter.

3.3 The Banaras School

There was a parallel development, albeit a few years later, in the campus of Banaras Hindu University, where, in 1932 Vishnu Vasudeva Narlikar took charge as Professor and the Head of Department of Mathematics. Narlikar had just returned from Cambridge University, where, as the Isaac Newton Scholar he was working under Eddington. Narlikar was in fact due to go back to Cambridge and later spend a year in Caltech working at the Mount Wilson Observatory and the Department of Astronomy. However, the founder of Banaras Hindu University, the venerable Pandit Madan Mohan Malaviya, always on the look out for talented Indian academics peruaded him to join Banaras Hindu University and continue his researches there while attracting good students to mathematics. Narlikar accepted the challenge and did not go back to Cambridge.

While in a reminiscent mood, Narlikar recalled to this author the episode when, in 1930–31, as a fresh research student, he solved the Einstein field equations for a homogeneous and isotropic universe, *dropping* Einstein's condition of a static model. He got a family of solutions of the expanding universe models which would be of interest to the small family of cosmologists and extragalactic astronomers. His guide Eddington was impressed by the work and offered to communicate it to the *Monthly Notices of the Royal Astronomical Society*. Just then he received in post a paper by Abbé Lemaitre in which the very same problem had been solved. The paper in fact published in 1927, was in a little known Belgian journal (Lemaitre 1927) and Eddington had been unaware of it. So he regretfully told Narlikar that his work could not be published. However, to bring to attention of the larger English speaking astronomy community, Eddington communicated Lemaitre's paper after translating it in English, to the same journal MNRAS. As luck would have it, even Lemaitre turned out to have been anticipated by Alexander Friedmann (1922, 1924)! Relativistic cosmology was, indeed, a 'hot' subject in the 1920s–1930s.

After joining Banaras Hindu University Narlikar concentrated on the different aspects of GR. He recalled later that he had asked S. Chandrasekhar (who, in the 1930s had worked on stellar models) if there was any scope for applying GR to stars. Chandrasekhar replied in the negative, thus discouraging Narlikar from pursuing that line of research. The applications potential of GR for massive dense objects thus remained untested for three decades.

Some of the work from the Banaras school relevant to relativistic astrophysics may be described as below.

In 1922 the noted mathematician, T.Y. Thomas had proved that in Riemannian manifold of 4 dimensions only 14 independent curvature invariants can be constructed. But the explicit construction of these 14 invariants using the curvature tensor and the Weyl tensor was first given by Narlikar and Karmarkar (1949). However, this work was published in the *Proceedings of the Indian Academy of Sciences*, a journal which did not have much circulation outside India. Unaware of this work therefore, several years later Geheniau and Debever in 1956 did the same work for which they have been given credit. This was noticed by A.R. Prasanna, a student of Narlikar at Pune who pointed out to Professor Geheniau this fact when they met in 1972, at the Dirac Symposium at Trieste. Geheniau readily agreed that these invariants should be called 'Narlikar-Karmarkar invariants'.

These invariants are important in deciding if a spacetime manifold has singularities. The question of singularities became relevant to reality by the discovery of collapsed massive objects in the form of quasars in 1963. Will the spacetime in the neighbourhood of such massive objects develop a singularity? If so how to spot it in a coordinate-invariant fashion? This is where the curvature invariants become important.

Work of a more mathematical nature came out of the studies of Narlikar and his students Ramji Tiwari and Kamala Prasad Singh. Tiwari was concerned with the unified field theory proposed by Einstein in the late 1940s and examined in detail the interaction between gravitation and electromagnetism (Narlikar and Tiwari 1949). Singh, on the other hand, worked on metric invariants. His work of the Christoffel symbols is of interest in bringing out the role of coordinate transformation that lead to indeterminateness (Narlikar and Singh 1951).

General relativity has the unique feature that it contains the equations of motion of the sources and the method of deriving them was indicated by Einstein, Infeld and Hoffmann (1938). Narlikar's student, B.R. Rao worked on the details of this problem and pointed out some corrections to the Einstein, Infeld and Hoffmann (EIH) work. This was recognized by Infeld and Hoffmann. The Narlikar-Rao paper (1955) appeared in print in the year that Einstein died.

But perhaps the most important paper to come out of the Banaras school was by P.C. Vaidya, one of the earliest students of V.V. Narlikar. We will describe it next.

3.4 The Vaidya Solution

P.C. Vaidya started his research career as a student of V.V. Narlikar. Himself a postgraduate of Bombay University, Vaidya enroled himself as an external research student of Narlikar in the Banaras Hindu University in the year 1942–43. Essentially living on his savings he made them stretch out to this period during which he also had to support a family of wife and child. Yet during those 2 years Vaidya was able to produce work that was to prove to be of very special interest to relativistic AP about 25 years later.

Basically the 'Vaidya solution' is a generalization of the Schwarzschild solution, the main difference between the two being that while the exterior of the gravitating sphere in the Schwarzschild solution is empty, the sphere in the Vaidya solution is radiating. Evidently, the situation described in the Vaidya solution is time-dependent; not static. We summarise this work below: for details of this solution see Vaidya (1943, 1950)

The 1950 paper quotes V.V. Narlikar (1939): 'If the principle of energy is to hold good, that is, combined energy of the matter and the field is to be conserved, the system must be an isolated system surrounded by flat spacetime. A spherical radiating mass would probably be surrounded by a finite and non-static envelope of radiation with radial symmetry. This would be surrounded by a radial field of gravitational energy becoming weaker and weaker as it runs away from the central body until at last the field is flat at infinity. It has to be seen whether and how this view of the distribution of energy is substantiated by the field equations of relativity.' This conjecture was borne out by the Vaidya solution.

To start with take the four spacetime coordinates to be $x^0 = t, x^1 = r, x^2 = \theta, x^3 = \phi$. A star of mass M and radius r_0 is supposed to start radiating at time t_0 and as time goes on, the zone of radiation increases in thickness, its outer surface at time $t = t_1 > t_0$ being given by $r = r_1 > r_0$. For $r_0 \leq r \leq r_1$ and $t_0 \leq t \leq t_1$, the line element is given by

$$ds^2 = e^{\nu}\, dt^2 - e^{\lambda}\, dr^2 - r^2\,(d\theta^2 + \sin^2\theta\, d\phi^2). \tag{3.4}$$

where both λ and ν are functions of r, t only. The outflowing radiation is described by the energy momentum tensor

$$T^{ik} = \rho v^i v^k, \tag{3.5}$$

where ρ is the density of radiation and v^i is the null vector representing its flow direction. For radial flow, we have $v^2 = v^3 = 0$. The field equations then give (with $G = 1, c = 1$),

$$e^{-\lambda} = 1 - \frac{2m}{r},\; m = m(r,t), \tag{3.6}$$

and

$$e^{\nu/2} = -\frac{\dot{m}}{m'}\left(1 - \frac{2m}{r}\right)^{-1/2}, \tag{3.7}$$

where m satisfies the relation

$$m'\left(1 - \frac{2m}{r}\right) = f(m). \tag{3.8}$$

The dot and dash denote differentiations with respect to t and r, respectively.

The function $f(m)$ is so far arbitrary but needs to be specified by the physical conditions that lead to the radiation from the star, whose mass m decreases at a rate determined by the amount of energy radiated by it. The radiation envelope of the star is described by the line element

$$ds^2 = \frac{\dot{m}^2}{f^2}\left(1 - \frac{2m}{r}\right) dt^2 - \left(1 - \frac{2m}{r}\right)^{-1} dr^2 - r^2\,(d\theta^2 + \sin^2\theta\, d\phi^2). \tag{3.9}$$

The energy conservation relation is described by the condition

$$\frac{dm}{d\tau} \equiv v^0\frac{\partial m}{\partial t} + v^1\frac{\partial m}{\partial r} = 0. \tag{3.10}$$

Of course, this equation is automatically satisfied if all field equations are satisfied. This formalism has been useful in the context of bright radiating objects in AP, such as quasars, active galactic nuclei, gamma ray bursts, etc.

3.5 The Raychaudhuri Equation

A major advance in the field of gravitational collapse came from the work of A.K. Raychaudhuri (1955). Before describing it we may mention that Raychaudhuri was essentially a loner when he did this work. Having studied in Calcutta, Raychaudhuri naturally started research in N.R. Sen's group. However, he soon discovered that he worked better in isolation. So, while at the Indian Association for Cultivation of Science, he turned his attention to the singularity problem. Raychaudhuri's work was originally concerned with relativistic cosmology but was later found to be applicable to finite massive distributions of matter as well. The aim was to find out if the presence of shear and rotation would prevent the gravitational collapse of such distributions. We will now briefly describe this work and its significance for relativistic AP.

The Raychaudhuri equation arises in relativistic cosmology when we study the bundle of timelike geodesics defined by the Weyl postulate. If u^i is the unit tangent to the geodesic, we define the spin-vorticity 3-tensor for the cosmic fluid by $\omega_{\mu\nu} = \frac{1}{2}[u_{\mu;\nu} - u_{\nu;\mu}]$. Writing the line element in the form

$$ds^2 = dt^2 + 2g_{0\mu}dtdx^\mu + g_{\mu\nu}dx^\mu dx^\nu, \tag{3.11}$$

where the geodesics are specified by x^μ = constant and t is the cosmic time, the (0,0) component of field equations in the case of dust of density ρ then becomes

$$\frac{\ddot{Q}}{Q} = \frac{1}{3}(2\omega^2 - 4\pi G\rho - \phi^2) \tag{3.12}$$

where $Q^6 = -g$ and

$$2\omega^2 = -g^{\lambda\mu}g^{\sigma\tau}\omega_{\lambda\sigma}\omega_{\mu\tau},$$

$$\phi^2 = \frac{1}{4}g^{\mu\nu}\dot{g}_{\nu\sigma}g^{\sigma\lambda}\dot{g}_{\lambda\mu} - \frac{1}{3}\left(\frac{\partial}{\partial t}\ln\sqrt{-g}\right)^2. \tag{3.13}$$

The ϕ term is identified with shear and it goes the opposite way (to the spin term) through promoting singularity by helping the scale of the cosmic volume, Q approach zero. It vanishes when the expansion is isotropic.

The Raychaudhuri equation can be stated in a slightly different form as a *focussing theorem*. In this form it describes the effect of gravity on a bundle of null geodesics spanning a finite cross-section. Denoting the cross-section by A, we write the equation of the surface spanning the geodesics as f = constant. With the normal to the cross-sectional surface being $k_i = \partial f/\partial x^i$. By the analogue of the hydrodynamic conservation law, we deduce

$$k^l A_{,l} = [k^l{}_{;l}]A. \tag{3.14}$$

Additionally we also have from the null geodesic condition

$$k^l k_{i;l} = 0. \tag{3.15}$$

Using a calculation similar to that which leads to the geodetic deviation equation, we now get the focussing equation as

$$\frac{1}{\sqrt{A}}\frac{d^2\sqrt{A}}{d\lambda^2} = \frac{1}{2}R_{im}k^i k^m - |\sigma|^2, \tag{3.16}$$

where

$$|\sigma|^2 = \frac{1}{2}k_{i;m}\,k^{i;m} - \frac{1}{4}[k^n_{;n}]^2. \tag{3.17}$$

The jvneq16 is similar to the Raychaudhuri equation with $|\sigma|^2$ being the square of the magnitude of shear. With Einstein's equations, we can rewrite jvneq16 as

$$\frac{1}{\sqrt{A}}\frac{d^2\sqrt{A}}{d\lambda^2} = -4\pi G(T_{im} - \frac{1}{2}g_{im}T)k^i k^m - |\sigma|^2. \tag{3.18}$$

For focussing of the bundle of rays we need $A \to 0$, so that the right hand side should be negative. This is helped by the shear term in the above equation just as Raychaudhuri had found. The first term on the right hand side of the focussing equation also has this property if

$$(T_{im} - \frac{1}{2}g_{im}T)k^i k^m \geq 0. \tag{3.19}$$

For dust we have $T_{im} = \rho u_i u_m$ and this condition is satisfied with the left hand side equalling $\rho(u_i k^i)^2$. (Remember that k_i is a null vector, so that $g_{im}k^i k^m = 0$.) Thus the normal tendency of matter is to focus light rays by gravity.

The *singularity* theorems of Penrose and Hawking (see Hawking and Ellis 1973) use this basic feature to state conditions that inevitably lead to spacetime singularity. The condition of the positivity of the T_{ik} term in equation (3.19) above plays a crucial role in general. We will not go into these details except to highlight it as a field deserving further research. In particular, the positive energy condition suggests that there may be non-singular spacetimes if it is violated and there are negative energy fields. We will now describe a line of thinking where such fields are used to avoid the initial (or *any*) singularity.

The Raychaudhuri equations for geodesic congruences has been around for more than half a century. In 1995, it was shown by Capovilla and Guven, that a generalization of these equations for families of extremal timelike membranes (D dimensional surfaces in a N dimensional background) is possible. Several illustrative examples and generalizations were worked out by Sayan Kar (1995, 1996a, 1996b). Later he also proved a focusing theorem for timelike surface congruences (Kar 1997).

Varun Sahni has also been interested in the singularity problem in higher dimensions. Recently Sahni and his colleagues have demonstrated that the braneworld cosmology, in which our 3+1 dimensional universe is a 'brane' embedded in a five dimensional 'bulk', has many important consequences (Sahni and Shtanov 2002). For instance, an expanding universe can encounter a new class of *quiescent* singularities at which the Hubble parameter and the density and pressure of matter remain finite, while the deceleration parameter and the Kretchmann invariant $R_{iklm}R^{iklm}$ diverge (Shtanov and Sahni 2002).

Another approach to singularities is through a generalization of the Hoyle-Narlikar concept of the *C*-field which has negative energy and stresses (see Hoyle and Narlikar 1962 for the first paper on this topic). In today's terminology such a scalar field is called a *phantom field.* M. Sami (Sami and Toporensky 2004) has worked in this area in relation to cosmology.

3.6 Naked Singularities

While the Penrose-Hawking theorems suggest the ubiquitousness of spacetime singularities, the question arises as to whether the singularity can be seen by an external observer in the case of a collapsing massive object. While the 'cosmic censorship hypothesis' suggests that the singularity will always be hidden by a black hole type horizon, some recent work by P.S. Joshi and his collaborators has turned up significant counter-examples.

A key result that emerges from these investigations is that visible ultra-dense regions or naked singularities arise naturally and generically as the final outcome of gravitational collapse in Einstein's gravity, in a wide variety of physically realistic situations. While it predicts their existence, the GR may no longer hold in these very late stages of collapse, and quantum gravity should take over to resolve the classical spacetime singularity. The quantum effects from a visible extreme gravity region could then propagate to outside observers to provide a possible laboratory to understand Planck scale physics and quantum gravity.

Joshi and Dwivedi (1993) carried out detailed analysis of naked singularities in dust collapse models to gain insights into the issue. Later this work was extended to models with pressure and also to anisotropic collapse. The question is of basic significance to black hole physics since the deductions like the second law of black hole physics are based on the validity of the cosmic censorship hypothesis. Details of such approaches may be found in Joshi's recent book (Joshi 2007). The implication of this type of work for the generally accepted behaviour of collapsing bodies is immense.

A. Banerji and collaborators (2003) studied the Tolman-Bondi type gravitational collapse in the spacetime with more than five dimensions and found that close to the centre in a marginally bound case naked singularity never appears. In another case Banerji et al. (2002) discussed a fluid sphere with outgoing heat flux and showed that trapping surface can be avoided till the singularity is attained at the end of the collapse. The non-occurrence of the horizon may be due to the mass-energy loss being balanced by the fall of the boundary radius.

3.7 Black Hole Astrophysics

The popularity of black holes amongst astrophysicists since the 1970s is reflected in the work done by Indian astrophysicists. The early work by Vishveshwara in 1970 on the quasi-normal modes (QNM) of oscillations of black holes is quite well cited. They offer a method of observing black holes directly. An important question that arises is in regard to the sensitivity of QNM to perturbing influences. This has been examined by Aguirregabiria and Vishveshwara (1996) using different equivalent potentials expected to be generated by the perturbations.

A phenomenal amount of research has been carried out on isolated black holes. Gravitational fields of such black holes are asymptotically flat and time independent. However, one cannot rule out black holes that are surrounded by matter distributions or imbedded in the cosmological background. Under these circumstances, the conditions of time symmetry and asymptotic flatness would have to be relaxed. A systematic study of such black holes shows how the properties of isolated black holes are significantly modified (Nayak et al. 2000; Ramachandra and Vishveshwara 2002, 2003; Ramachandra et al. 2003).

In the context of accretion dynamics A.R. Prasanna studied the stability of the accretion disc around compact objects with or without associated electromagnetic fields. In this context Abramowicz and Prasanna (1990) investigated the inertial forces very close to a black hole and found that there is the possibility of reversal of the centrifugal force in this region. This could affect the overall viscous drag and inwards advection during accretion. This in turn could influence the fluid flow towards the last stable circular orbit around the compact object.

N.K. Dadhich and collaborators revisited the Penrose Process (PP) of extracting energy from a black hole. The original PP had the unrealistic demand that the fragment of the original piece of matter falling into a negative energy orbit had to be accelerated to half the speed of light. In the absence of any mechanism for doing so, the PP became rather esoteric. Dadhich showed that if there is (electro)magnetic field around inside the black hole ergosphere, that could do the job of sending the piece into the required negative energy orbit, without having to accelerate it unrealistically. This modified PP is known as the 'Magnetic Penrose Process' and it has been described in an article by Dadhich (1989).

Jishnu and Mira Dey and colleagues (see Bagchi et al. 2008) have provided a rather exotic explanation of the afterglow seen after a gamma ray burst (GRB). They invoke a strange star which collapses into a black hole depending on its mass and temperature. A black hole sucks in matter and in the process creates an outward jet causing GRB. The ejected material in the GRB gives rise to afterglows that bear the signature of the GRB.

In a significant work, A. Banerji and S.B. Duttachoudhury (1989) studied the boundary conditions at the surface of a charged viscous fluid sphere with outgoing heat and radiation flux. Thus the exterior solution is given by the Reissner-Nordstrom-Vaidya metric and the Israel-Darmois boundary conditions are used. In Romesh Kaul's work the Chern-Simons theory finds application in the study of black holes in quantum geometric formulation of gravity. In fact, $SU(2)$ Chern-Simons theory represents the boundary degrees of freedom of a non-rotating black hole in four dimensions. An exact formula for the dimensionality of the Hilbert space of the boundary states of $SU(2)$ Chern-Simons theory has been derived. This provides a way of calculating entropy of a four dimensional Schwarzschild black hole. While the formula obtained gives an *exact* quantum entropy for any size of the area, in the limit of large areas, the result does indeed agree with the semi-classical Bekenstein-Hawking entropy proportional to the horizon area A. The next order term for large areas is lnA with a definite coefficient, $-3/2$ (Kaul and Majumdar 2000).

After this pioneering result was obtained, some other black holes have been shown to posses this log (horizon area) correction to their entropy with the same coefficient $-3/2$. This includes those in string theories

(Kaul 2003). Like the Bekenstein-Hawking formula, this correction to the black hole entropy appears to be universal as argued by Carlip (2000).

This work is typical of the interest amongst particle physicists in the notion of a black hole viewed in the framework of quantum theory. Although talking about black holes, the ideas developed following Stephen Hawking's pioneering work, have little common ground with the black holes discussed by astrophysicists. As our interest in this chapter is limited to relativistic AP, we will have to skip the very productive work in quantum fields in curved spacetime, quantum gravity, *per se*, string theory, etc.

3.8 White Holes

The simplest way of describing a white hole is as a time reversed version of a black hole. Thus instead of gravitational collapse of a massive object ending in spacetime singularity, one is talking here of an object emerging from a spacetime singularity with high outward velocities. In a sense the Big Bang universe is the supreme example of a white hole. How will a white hole appear to an external observer?

In the mid-1970s J.V. Narlikar and his collaborators (see Narlikar et al. (1974); Narlikar and Apparao (1975)) showed that initially there will be high blueshift of light emitted by the white hole and that as the expansion proceeds the magnitude of the blueshift will decline and ultimately the shift will be towards the red end. The solution of the white hole emission problem is exact if the white hole is made of dust (i.e. it is pressure-free).

Somewhat similar work was done by A. Banerji (1966) showing that the frequency shift observed by an external observer is a mixture of Doppler shift and gravitational redshift.

D.M. Eardley (1974) had argued in a short paper that a white hole is unstable and gets converted into a black hole because of strong accretion near the horizon as it tries to emerge from it. T.K. Dey and S. Banerji (1991, 1993) have shown, however, that a white hole may or may not convert to a black hole as it encounters a collapsing spherical dust shell, depending on the initial conditions.

3.9 Gravitational Lenses

Although 1979 saw considerable publicity being given to gravitational lensing in quasar astronomy, there had been earlier workers in the field also. Thus one may start with Fritz Zwicky (1937), Refsdal (1964) and Barnothy (1965). A direct application of gravitational lensing to quasar observations was suggested first by J.V. Narlikar and S.M. Chitre (1978) who argued that the apparent superluminal separation of VLBI sources in a quasar may have been magnified. Thus a subluminal separation may be seen as superluminal because of lensing by an intermediate galaxy.

After 1979 and the realization that the so-called twin quasars may be the two lensed images of a single source, there were several attempts at modelling the suspected lensed cases. In this field Indians have played a lead role and the main players in this field have been S.M. Chitre, K. Subrahmanian, D. Narasimha and S. Nair. General relativistic bending of light is invoked in constructing these models.

In Einstein's general theory of relativity, light rays can get bent by the gravitational influence of mass distributions. This effect called gravitational lensing can lead to several interesting phenomena. One of the most interesting consequences of gravitational lensing is the multiple imaging of a distant quasar by an intervening galaxy. The first case of such multiple imaging was discovered in 1979 by Walsh, Carswell and Weymann. Optical observations of two quasars 0957+561 A and B, separated by about 6 arc seconds, showed almost identical spectra. This was surmised to be one quasar whose light has been bent by an intervening galaxy such that it arrives along two different paths. This discovery of multiply imaged quasars led to an explosion of interest in gravitational lensing. Subsequent to the discovery of the twin quasar, detailed lens models for a large number of observed cases of multiple imaging were constructed by Indian astrophysicists (Narasimha, Subramanian and Chitre 1982, 1984a, 1984b; Subrahmanyan et al. 1990; Nair, Narasimha and Rao 1993; Nair and Garrett 1997; Nair 1998). The gravitational lens models probed the general properties of the intervening galaxies and clusters. In the process the Indian group developed one of the first numerical codes to extensively model cases of such strong lensing.

Stars in the lensing galaxy can also further amplify the light from the background quasar. Since the stars move in the galactic potential, this amplification can vary on timescale of a few years. This phenomena called microlensing was also extensively studied (Narasimha, Subramanian and Chitre 1984a; Nityananda and Ostriker 1984; Subramanian, Narasimha and Chitre 1985), and even used to probe the central black holes in lens galaxies (Narasimha, Subramanian and Chitre 1986). If the source which is being microlensed is moving relativistically then fast variations can result even in radio wavelengths (Gopal-Krishna and Subramanian 1991).

By the end of the 1980s a new phenomenon was discovered, the lensing of extended sources like galaxies, by intervening massive galaxy clusters, to produce giant arcs (Lynds and Petrosian 1986, 1989). These were also modeled by Narasimha and Chitre (1988, 1993) and used to probe the matter distribution in clusters. Gravitational lensing has proved to be one of the most important ways in which to probe the dark matter distribution of galaxy clusters.

A very interesting case of multiple imaging was also discovered by Indian astronomers (Subrahmanyan and Rao 1988), first through radio observations. This lens (PKS 1830-211) called by some as the "Ooty lens" was modeled extensively by Subrahmanyan et al. (1990) and Nair, Narasimha and Rao (1993), was later confirmed to be an example of an Einstein ring.

Another important advance, made by Nityananda (1990a, 1990b) and Blandford and Narayan (1986), was to formulate the gravitational lensing equations as an application of Fermat's principle. General theorems on the conditions for multiple imaging by a smooth, bounded gravitational lens were also proved by Subramanian and Cowling (1986). Padmanabhan and Subramanian (1988), extended these considerations to cosmologically extended thick gravitational lenses.

3.10 Gravitational Waves

In the last decade or so Indians have begun making their presence felt amongst the small international community of workers in the field of gravitational waves (GWs). On the instrumentation side, the detectors need to meet very demanding criteria for dealing with low signal to noise ratio. On the software side frontier level work is going on in the area of data analysis. S.V. Dhurandhar has played a lead role in this field.

The detectors are optimally looking at cosmic sources in the form of inspiralling compact binaries; for inspiralling compact (neutron stars, black holes) binaries are considered to be the most promising sources for ground-based detectors. In two papers, the foundation of a computationally optimal scheme in searching for inspiralling waveforms was presented. Further, efficient hierarchical methods in searching for inspirals were developed which reduces the computational cost several times has become the standard reference for designing numerical codes. Another important contribution is coherently extracting inspiralling binary signals from a network of detectors. This work enabled Inter-University Center for Astronomy and Astrophysics (IUCAA) to join the LIGO Science Collaboration (LSC) which is the most important worldwide collaboration and will run the US-based LIGO detectors. The coherent search takes into account the phase information into building up the network statistic and therefore is superior in performance to the coincidence search where each detector is treated in isolation. A quantitative comparison of the two strategies is being performed currently with the Japanese.

Recent work comprises the radiometric search for stochastic GWs and also for periodic sources such as pulsars or rotating neutron stars. This work is within the LSC collaboration with Caltech and Albert Einstein Institute, Potsdam, Germany. The radiometric search has the advantage that the search is possible in the sensitive frequency band of GW detectors. This is a directed search involving a network of detectors.

In LISA (Laser Interferometric Space Antenna) data analysis, an important problem is cancellation of laser frequency noise which arises because of the impossibility of LISA to maintain equal armlengths. The cancellation is achieved with a scheme known as time-delay interferometry (TDI). A rigorous mathematical foundation of this scheme was laid out with the use of Gröbner basis methods and modules over polynomial rings. The TDI combinations form the first module of syzygies well known in commutative algebra and algebraic geometry. In the case of stationary LISA, its generators were obtained so that all TDI combinations could be available in principle. Work is in progress when the armlengths change with time. This leads to non-commutative operators. One way is to optimise the spacecraft orbits so that the rate of change of armlengths is acceptably small and the other approach is to generalise the previous results to the non-commutative case. This work is in collaboration with the French.

As mentioned above, inspiralling compact binaries comprising neutron stars and black holes are the most promising sources for ground-based laser interferometric GW detectors like LIGO and Virgo and space-based detectors like LISA. Data analysis issues must therefore be supplemented by theoretical analysis of these sources. Their detection and parameter estimation employs matched filtering and this requires the construction of effectual, faithful and efficient template banks. The theoretical input for building template banks are high accuracy phasing formulas for the evolution of the GW phase under gravitational radiation reaction far beyond the leading radiation reaction at 2.5 post-Newtonian (PN) order used in the timing analysis of the binary pulsar. Starting with the 2PN generation of GW (providing the 4.5PN terms in the equations of motion and forming the starting point for all GW data analysis in LIGO and Virgo), over a decade the multipolar post-Minkowskian formalism matched to a PN source has been systematically extended to 3.5PN order. For many years the 3PN results (Blanchet et al. 2002, Blanchet and Iyer 2005) remained partial due to the presense of undetermined parameters arising from the incompleteness of the Hadamard regularization used to deal with the divergent self-field effects. Recently, by the use of a more powerful dimensional regularization these 3PN parameters have been determined (Blanchet et al. 2004) completing the GW phasing to 3.5PN. Further, the full gravitational waveform generated by inspiralling compact binaries moving in quasi-circular orbits is computed at the 2.5PN and 3PN approximation (Blanchet et al. 2008). This 3PN amplitude accurate and 3.5PN phase accurate phase waveform will form the basis for searching and deciphering the GW signals in the current and future network of GW detectors and also for calibrating and interpreting the recent exciting results of numerically generated waveforms for the merger and ringdown of binary black holes (Blanchet et al. 2004). More generally, for quasi-elliptic orbits the formalism can describe both their secular evolution under 3PN gravitational radiation reaction (Arun et al. 2008) and the smaller but fast orbital scale periodic terms (Damour et al. 2004). Resummation techniques like Pade approximants and effective one body methods have been used to extend the validity of standard PN approximants (Damour et al. 1998) and template families more critically classified for comparison (Damour et al. 2001).

The implications of the full waveform that brings into play harmonics beyond the dominant have been explored for LISA. By inclusion of higher harmonics in matched filters more massive systems that were previously thought to be *not* visible in LISA are detectable with reasonable SNRs. Moreover, the angular resolution of LISA increases by more than a factor of 10 thereby making it possible for LISA to identify the host galaxy/galaxy cluster. Thus, LISA's observation of certain binary supermassive black hole (SMBH) coalescence events could constrain the dark energy equation of state to within a few percent, comparable to the level expected from other dark energy missions (Arun et al. 2007). Finally, LISA provides a unique opportunity to probe the non-linear structure of PN theory both in the context of GR and its alternatives (Arun et al. 2006).

3.11 The Present GR-AP Infrastructure in India

I have briefly described the range of subjects covered in India in the field of GR and AP. Additionally work is being done in such frontier areas like quantum gravity via loops (G. Date) as well as via strings, alternative models of the universe (J.V. Narlikar), quantum fields in curved spacetime (Padmanabhan, Sriramkumar), multi-dimentional cosmologies (M. Sami, V. Sahni), and the mathematical/geometrical aspects of GR and other gravity theories (Pankaj Joshi). I have mentioned the names of a few leading workers in the parentheses after each topic. It will take me well beyond the scope of this chapter to touch those areas.

It is worth recording here that Indians have been regularly participating in the international essay competition conducted by Gravity Research Foundation of USA. Several honourable mentions have been won as well as prizes. T. Padmanabhan has received prizes four times in this contest during the last 7 years including the first prize in 2008. This is the first time work done in India (Padmanabhan 2008) has received the top award from Gravity Research Foundation.

I end this chapter brief review of where in India important centres of GR-AP exist. The Tata Institute of Fundamental Research, Mumbai has scientists in this field working largely in Astronomy Department. Another group exists in the Raman Research Institute, Bengaluru and the Indian Institute of Astrophysics, Bengaluru. The former of the two is more theoretical while the latter is inclined towards AP. Then we have the Physical Research Laboratory, Ahmedabad and the Institute of Physics, Bhuvaneswar with a few workers, again inclined towards AP in the former and GR in the latter. More mathematically oriented work is being done in the Institute of Mathematical Sciences, Chennai and the Harishchandra Research Institute in Allahabad.

These places, as will be noticed, are all *outside* the University Sector, in contrast to the early days when work was done mainly in the universities like Calcutta and Banaras. The flagship for the University Sector in this repect is the IUCAA in Pune which has strong research base in both GR and AP. Smaller groups do exist in the universities, for example, in Kolkata there are the Calcutta University and Jadavpur University both of which have research in GR-AP going on. A new Centre for Theoretical Physics has been opened at the Jamia Millia Islamia in Delhi with GR as the thrust area. Thanks to IUCAA the pedagogical activities in GR-AP have increased considerably, ranging from introductory summer schools to advanced specialist workshops.

Two national organizations look after the interests of GR and AP. The Indian Association of General Relativity and Gravitation (IAGRG) was founded in 1969, whereas the Astronomical Society of India started in 1973. Both have many common members and each society meets once in about 18 months. Besides, several international meets have been successfully held in India, including the General Assembly of the IAU in Delhi in 1985, the International Conference on Gravitation and Cosmology in Goa in 1987, the Asia Pacific Regional Meeting of the IAU in 1993, the GR-15, of the International Society of General Relativity and Gravitation in 1997. The last two were hosted by IUCAA.

Given these data, the situation looks brighter for the future, except for the caveat: Can we continue to attract young talent to GR-AP? India's participation in major international facilities like GW detectors, ultra large telescopes, large particle accelerators, etc. will help towards this goal.

References

1. Abramowicz MA and Prasanna AR MNRAS, 245: 720 (1990)
2. Aquirregabiria JM and Vishveshwara CV. Phys. Lett. A, 210, 251 (1996)
3. Arun KG. Blanchet, Luc, Iyer, Bala R and Qusailah, Moh'd SS. Phys. Rev. D, 77, 064035 (2008)
4. Arun KG, Iyer, Bala R, Sathyaprakash BS, Sinha, Siddhartha and Broeck, Chris Van den: Phys. Rev. D., 76, 104016 (2007)
5. Arun K, Iyer BR, Qusailah MSS and Sathyaprakash BS. *Class. Quant. Grav.*, **23**, L37 (2006)
6. Bagchi M. et al. *MNRAS*, **387**, 115 (2008)
7. Banerji A. *Phys. Rev.*, **150**,1086 (1966)
8. Banerji A and Duttachoudhury SB. *GRG*, **21**,785 (1989)
9. Banerji A. and Chakraborty D.S. *IJMP*, **12**,1255 (2003)
10. Banerji A. Chatterjee S, and Dadhich NK. *MPL*, **A17**, 2335 (2003)
11. Barnothy JM. *A.J.*, **70**, 666(1965)
12. Blanchet Luc, Iyer, BR and Joguet B. *Phys. Rev D.*, **65**, 064005 (2002)
13. Blanchet L and Iyer BR. , *Phys. Rev. D*, **71** , 024004 (2005)
14. Blanchet L, Damour T. Esposito-Farèse, G. and Iyer, B.R. *Phys. Rev. Lett.*, **93**, 091101 (2004)
15. Blanchet, Luc, Faye, Guillaume, Iyer, Bala R, and Sinha Siddhartha , *Class. Quant. Grav.*, **25**, 165003 (2008)
16. Blandford R and Narayan R. , *Astrophysical Journal*, **310**, 568-582 (1986)
17. Carlip, S. *Class.Quant. Grav.* **17**, 4175 (2000)
18. Capovilla, and Guven, (1995)
19. Chitre SM and Narlikar JV. *MNRAS*, **187**, 655 (1978)
20. Damour T, Gopakumar A and Iyer BR. *Phys. Rev.*, **D**, **70**, 064028 (2004)
21. Damour T, Iyer BR, and Sathyaprakash BS. *Phys. Rev. D.*, **57**, 885 (1998)
22. Damour T, Iyer BR, and Sathyaprakash BS. *Phys. Rev D.*, **63**, 044023 (2001)
23. Datt B. *Z. Phys.*, **108**, 314 (1938)
24. Dadhich NK. *Phys. Reps* (1989)
25. Dey TK and Banerji S. *Phys. Rev.*, **D44**, 325 (1991)
26. Dey TK and Banerji S. *Phys. Rev.*, **D48**, 3478 (1993)
27. Dhurandhar SV and Sathyaprakash BS. *Phys. Rev. D*, **49**, 1707 (1994)
28. Dhurandhar SV, Nayak R, and Vinet JY *Phys. Rev. D.*, **65**, 102002 (2002)
29. Dhurandhar SV, Vinet J-Y, and Nayak, KR. *Class. and Quantum Grav.*, **25**, 245002 (2008)
30. Eardley DM. *Phys. Rev. Letts*, **33**, 442 (1974)
31. Einstein A. *Preuss. Akad. Wiss. Berlin, Sitzber.*, 778, 799, 844 (1915)
32. Einstein A. *Preuss. Akad. Wiss. Berlin, Sitzber.*, 142 (1917)
33. Einstein A, Infeld L. and Hoffmann B. *Ann. Math.*, **39**, 65 (1938)
34. Friedmann A. *Z. Phys.*, **10**,377, (1924) *Z. Phys*, **21**, 326 (1922)
35. Geheniau J, and Debever R. *Bull. Acad. Belg.*, **42**, 114 (1956)
36. Hawking SW, and Ellis GFR. *The Large Scale Structure of Spacetime*, Cambridge (1973)
37. Hoyle F, and Narlikar JV. *Proc. R. Soc.*, **A**,(1962)
38. Joshi PS. *Gravitational Collapse and Spacetime Singularities*, (Cambridge University Press, Cambridge) (2007)
39. Joshi PS, and Dwivwdi IH. *Phys. Rev.*, **D47**, 5357 (1993)
40. Kar S. *Phys. Rev.*, **D52**, 2036 (1995)
41. Kar S. *Phys. Rev.*, **D53**, 2071 (1996a)
42. Kar S. *Phys. Rev.*, **D54**, 6408 (1996b)
43. Kar S. *Phys. Rev.*, **D55**, 7921 (1997)
44. Kaul RK. *Phys. Rev.*, **D68**, 024026 (2003)
45. Kaul RK, and P Majumdar. gr-qc/0002040, *Phys. Rev. Letts.*, 84, 5255-5257(2000)
46. Krishna Gopal, and Subramanian K. *Nature*, **349**, 766-768 (1991)

47. Lemaitre, Abbé G. *Annales de la Soc. Sci.de Bruxelles*, **XLVII A**, 49 (1927)
48. Lynds R, and Petrosian V. *BAAS*, **18**, 1014 (1986)
49. Lynds R, and Petrosian V. *Astrophys. J.*, **336**, 1-8 (1989)
50. Mitra S, Dhurandhar S, Souradeep T, Lazzarini A, Mandic V, Bose S, and Ballmer S. *Phys. Rev. D.*, **77**, 042002 (2008)
51. Nair S, Narasimha D, and Rao, AP. *Astrophys. J.*, **407**, 46-59 (1993)
52. Nair S, Garrett M. *Mon. Not. R. Astr. Soc.*, **284**, 58-72 (1997)
53. Nair S. *MNRAS*, **301**, 315-322 (1998)
54. Narasimha D, and Chitre SM. *Astrophys. J.*, **332**, 75 (1988)
55. Narasimha D. and Chitre SM. *Astron. and Astrophys.*, **280**, 57-62 (1993)
56. Narasimha D, Subramanian K, and Chitre SM. , *MNRAS*, **200**, 941-950 (1982)
57. Narasimha D, Subramanian K and Chitre SM. *MNRAS*, **210**, 79-88 (1948a)
58. Narasimha D, Subramanian K, and SM Chitre. *Astrophys. J.*, **283**, 512-514 (1948b)
59. Narasimha D, Subramanian K, and Chitre SM. *Nature*, **321**, 45-46 (1986)
60. Narlikar JV, Apparao MVK, and Dadhich NK. *Nature*, **251**, 590 (1974)
61. Narlikar JV, Apparao MK. *Ap.SS*, **35**, 321 (1975)
62. Narlikar VV. *Bombay Univ. J.*, **8**, 31 (1939)
63. Narlikar VV, and Karmarkar KR. *Proc. Ind. Acad. Sci.*, **29**, 91 (1949)
64. Narlikar VV and Rao BR. *Proc. Nat. Inst. Sci*, **A21**, 416 (1955)
65. Narlikar VV and singh KP. *Proc. Nat. Inst. Sci.*, **17**, 311 (1951)
66. Narlikar VV and Tiwari Ramji. *Phys. Rev. Ser.2*, **76**, 868 (1949)
67. Nayak KR, MacCallum MAH and Vishveshwara CV. *Phys. Rev. D.*, **63**, 024020 (2000)
68. Nityananda R. in: Gravitational lensing, Berlin and New York, Springer-Verlag,3-12 (1990a)
69. Nityananda R. *Current Science*, **59**, 1044 (1990b)
70. Nityananda R and Ostriker JP. *Jour. Astrophys. and Astron.*, **5**, 235-250 (1984)
71. Oppenheimer JR and Snyder H. *Phys. Rev.*, **56**, 455 (1939)
72. Padmanabhan T and Subramanian K. *MNRAS*, **233**, 265-284 (1988)
73. Padmanabhan T (2008) Gen. Rel. Grav. **40**, 2031 (2008)
74. Pai A, Dhurandhar SV, and Bose S. *Phys. Rev. D.*, **64**, 042004 (2001)
75. Ramachandra BS, and Vishveshwara CV. *Class. Quant. Grav.*, **19**, 127 (2002)
76. Ramachandra BS, and Vishveshwara CV. *Class. Quant. Grav.*, **20**, 5253 (2003)
77. Ramachandra BS, Nayak KR and Vishveshwara CV. *Gen. Rel. and Grav.*, **37**, 1977 (2003)
78. Raychaudhuri AK. *Phys. Rev.*, **98**, 1123 (1955)
79. Refsdal S. *MNRAS*, **128**, 295 (1964)
80. Sahni V and Shtanov Y. *JCAP*, **11**, 14 (2002)
81. Sami M and Toporensky A. *Mod. Phys. Lett.*, **A19**, 1509 (2004)
82. Sathyaprakash BS and Dhurandhar SV. *Phys. Rev. D*, **44**, 3819 (1991)
83. Sen NR. *Annls Phys.*, **4**, 73 (1924)
84. Sen NR. *Z. Phys.*, **40**, 667 (1927)
85. Sen NR. *Z. Astrophys.*, **7**, 188 (1933)
86. Sen NR. *Z. Astrophys.*, **9**, 215 (1934)
87. Sen NR. *Z. Astrophys.*, **9**, 315 (1935a)
88. Sen NR. *Z. Astrophys.*, **10**, 29 (1935b)
89. Sen NR. *Z. Astrophys.*, **14**, 157 (1937)
90. Sen NR. *Ind. J. Phys.*, **15**, 209 (1941)
91. Sen NR. and von Laue, M. *Annls Phys.*, **4**, 1882 (1924)
92. Sen NR and Roy TC. *Z. Astrophys.*, **34**, 84 (1924)
93. Shtanov Y and Sahni V. *Clss.Q. Grav.*, **19**, L101 (2002)
94. Subramanian K, Chitre SM and Narasimha D. (1985), *Astrophys. J.*, **289**, 37-51 (1985)
95. Subramanian K. and Cowling SA. *MNRAS*, **219**, 333-346 (1986)
96. Subrahmanyan R. and Rao AP. *MNRAS*, **231**, 229 (1988)
97. Subrahmanyan K, Narasimha D, Rao AP and Swarup G. *MNRAS*, **246**, 263 (1990)
98. Vaidya PC. *Curr. Sci.*, **12**, 183 (1943)
99. Vaidya PC. (1950)
100. Walsh D, Carswell RF and Weymann RJ. *Nature*, **279**, 381-384 (1979)
101. Zwicky F. *Phys. Rev.*, **51**, 679 (1937)

Chapter 4
Developments of Space Astronomy in India

B.V. Sreekantan

4.1 Introduction

The universe is pouring out information on its extent, structure, composition, variety of objects, their relative motions, their temperatures, time variations, the electric and magnetic fields in the various regions, the physicochemical and biological evolutionary chains and a host of other details, in the form of a variety of particulate and electromagnetic, radiations, neutrino and perhaps exotic particles yet unidentified, and has left it to the ingenuity of man to record these radiations and come out with a consistent, meaningful understanding of the origin, the fundamental constituents and the forces behind the vast range of phenomena witnessed. In this endeavour naked eye astronomy was naturally the oldest dating back to several millennia, followed by telescopic observation starting with Galileo in the 16th century, spectroscopic observations of celestial objects planets and stars in the 19th century. The installation of the large telescopes at Mount Wilson and Mount Palomar in California were landmark events in the long history of optical astronomy. A vast amount of data has been collected and analysed and very many intricate aspects of the universe have been figured out on the basis of optical astronomy alone eventhough it is based on an extremely narrow band of the electromagnetic spectrum. In the 1930s, serendipitously a new window of astronomy opened up radio astronomy which has supplemented and complemented optical astronomy through the discovery of a vast range of phenomena not seen in optical astronomy. A flavour of the type of new results from radio astronomy window is available in the accompanying chapter by Prof. Swarup.

While it was expected that lot more information must be available in the other bands of the electromagnetic spectrum infrared, ultraviolet, X-ray, gamma-ray, etc. observations in these bands had to await the development of space technology since the specially designed telescopes had to be carried to great heights in the atmosphere for reasons which become apparent from Figure 4.1 which shows the absorptions characteristics of the earth's atmosphere for the various bands of the spectrum. The space vehicles namely stratospheric balloons, rockets became available only in 1960s though they had been developed much earlier for military purposes. Also special types of radiation detectors, telemetry, telecommand systems were also needed in addition to computers in the ground sector for detailed analysis.

4.2 X-ray Astronomy

Observations with ground-based infrared telescopes designed to pick up radiations available in the infrared at the ground level through, the leakage in some of the windows (2–2 microns, etc.) and also the observation of ultraviolet radiation from the sun picked up at high mountain altitudes had given indication of bright prospects in astronomy in these bands at higher altitudes and naturally for space-based telescopes these bands were the first priority. However what came as a big surprise was the serendipitous discovery of a bright X-ray since in 1962 from an experiment which was designed for an entirely different purpose and opened up a rich field of space astronomy that had tremendous influence and impact on the other windows of astronomy and astrophysics and cosmology over the last four decades.

National Institute of Advanced Studies, Indian Institute of Science, Bengaluru - 560 012. e-mail: bvsreekantan@gmail.com

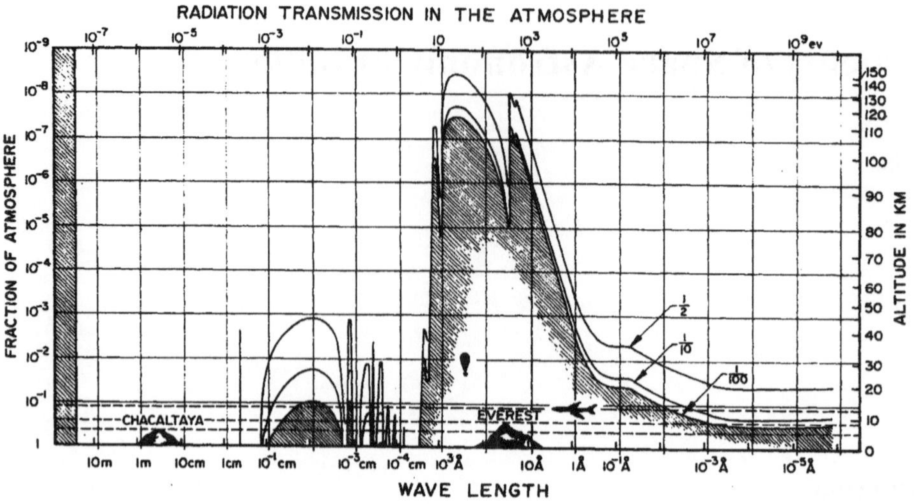

Fig. 4.1 The absorption characteristics of the atmosphere as a function of the wavelength (energy) of the electromagnetic radiation

The discovery of the first X-ray source ScoX-1 is has a very interesting story. Immediately after the second world war, a series of space probes were launched by the United States for exploring the interplanetary space. These led to the discovery that in addition to the enormous amount of heat and light emitted by the sun, there was also outpouring of a constant stream of particles comprising protons, helium nuclei and other high energy radiations. The famous cosmic ray physicist Bruno Rossi of Massachusetts Institute of Technology (MIT) realised that a consequence of this solar wind radiations impact on the moon would be to give rise to fluorescent X-rays. To test this conjecture, the MIT and AS&E group (R Giacconi et al. 1962). launched a rocket experiment in June 1962 with a simple payload of two parallely mounted thin window Geiger counters on the rotating rocket to record the fluorescent X-ray and adjusted the timing of the launch such that when the rocket was at high level, the moon would be in the horizon. If the hypothesis was correct then the peak in the counting rate as a function of azimuth would coincide with the direction of the moon. The analysis of the counting rate revealed that there was an increase in the general direction of the moon, but the peak was shifted by several degrees in azimuth with respect to the moon (Figure 4.2). It so happened that in those few minutes on that particular day, when the X-ray counters were scanning the horizontal sky one of the brightest X-ray source was sitting just next to the moon in angular direction and revealed itself. Confirmation came from a second experiment, carried out by the same group a few month later.

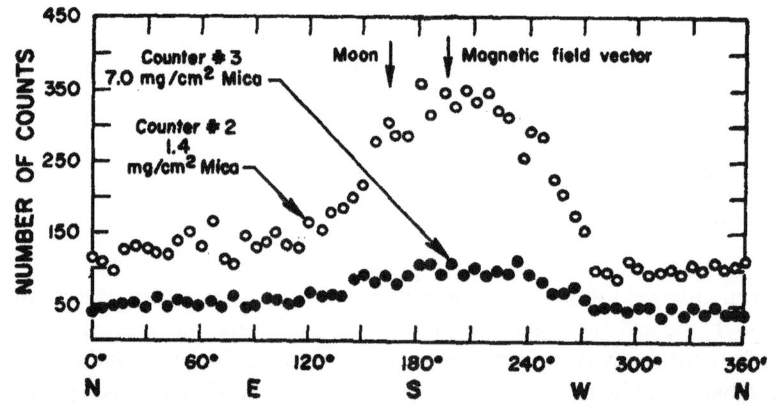

Fig. 4.2 Azimathal distributions of recorded counts from Geiger counters flown during June 1962 (R Giacconi, H Gursky, F Paolini, BB Rossi 1963, Phys Rev Lett 9:442)

What was most surprising about this X-ray source was its very high intensity. As early as 1949, an attempt was made to record X-rays from the sun by Burnight (1949)[7] using covered X-ray photographic plates flown to an altitude of 96 km without any success. In the late 50s Purcell and collaborators (1949) of Naval Research Laboratories set an upper limit of 10^{-8} ergs cm^{-2} sec^{-2}. The flux from the X-ray source ScoX-1 detected by the MIT and AS&E group was 10–100 million times that of the quiet sun in X-rays. A few months later George Clark (Purcell et al. 1949) of MIT discovered the emission of hard X-rays from the Crab Nebula in a balloon flight experiment. These two sources as X-ray sources were further confirmed by rocket experiments by the NRL group (Bowyer et al. 1964) in 1963. Since then a variety of experiments have been carried out with balloons, rockets and satellites by groups in the USA, UK, Soviet Union, Japan and India and several hundred X-ray sources have been discovered.

4.3 Development of X-ray Astronomy in India

It so happened that during the period 1965–1967, the author (BVS) was on sabbatical at MIT and had the wonderful opportunity of participating in the MIT and AS&E collaboration experiments in X-ray astronomy with rockets. The objective of the experiments that were being planned was to determine the angular sizes of the sources ScoX-1 and TauX-1 using a modulation collimator a novel idea that had been proposed by one of the Japanese colleagues Prof. Minore Oda who was also on sabbatical at MIT from the university of Tokyo. Though the design, fabrication, alignment of the collimator posed serious problems, these were overcome (Bradt et al. 1968) and the group had a successful rocket flight on March 8, 1966 which led to a measurement of the location of ScoX-1 to an accuracy of 1 arc minute and the angular size of the source was set at <20 arc seconds (Gursky et al. 1966). Further, with the accurate of position determination it became possible to identify with the help of optical astronomers at Palomar, the source ScoX-1 with a 12.5 magnitude star with an ultraviolet excess relative to normal stars. It was also found that the object was highly unstable in its continuum radiation.

In the same flight on March 8, 1966, the Crab Nebula was also observed. It was found that the emission region in the Crab was of finite size (Oda et al. 1967) and agreed with the earlier observation of Bowyer et al. (1964)[5] using the lunar occultation technique.

Rich with experience of working in this newly developing field of X-ray astronomy at MIT, I returned to Tata Institute of Fundamental Research (TIFR) in 1967. I realised that the high altitude studies group of TIFR which had been conducting experiments on primary cosmic rays and electrons with electronic and nuclear emission detectors, had all the expertise to undertake a vigorous programme in the field of hard X-ray astronomy using stratospheric balloons.

The hard X-ray balloon astronomy programme of TIFR started in Hyderabad from1968 itself just 6 years after the discovery of the first X-ray source ScoX-1. The Physical Research Laboratory group of Ahmedabad also statrted their programme in X-ray astronomy soon using the balloon facility at Hyderabad. One of the advantages of Hyderabad was at this latitude, most of the X-ray sources that had been discovered by then ScoX-1, CygX-1, TauX-1, CygX-3 and HerX-1 were available for observation at high enough altitude for convenient observation with X-ray detectors mounted on oriented platforms. However, one disadvantage of equatorial latitudes was that the tropopause is at a lower altitude and the temperature is quite low with the result that the balloon bursts due to brittleness of the fabric during night flights. The fabric manufactured in India was not suitable for night flights. We had to import special balloons for experiments in X-ray astronomy which most often required night flights. The very first source that was looked at by us is SoX-1 itself. This experiment Agrawal et. al. (1971) revealed the flattening of the spectrum beyond 40 KeV (Figure 4.3). Subsequent experiments with UHURU and other X-ray satellites have discovered a very large number of X-ray sources with rather unexpected properties not observed in optical and radio astronomies. X-ray telescopes with the detector systems like proportional counters and sodium iodide scintillation crystals had one advantage compared to the earlier recordings in optical and radio astronomy namely they could record the variations in intensity of the sources with time resolutions of micro and nanoseconds. This led to the discovery that the X-ray sources, most of them showed variations in intensity over short periods of time and also registered bursting activity in many of them. ScoX-1 was found to be one such with our balloon experiments from Hyderabad. As already pointed our the optical counterpart of ScoX-1 was a 12.5 magnitude star. When this was examined by telescopes attached with spectroscopes some new exciting results were found. Doppler shift of the spectral lines as a function of time together with time variations in X-ray intensity suggested a binary structure for this source with a period of 0.787 days. Also simultaneous optical and X-ray variations without any delay between signals have been interpreted in terms of optical emission arising from the bombardment of X-rays on the matter of accretion disc close to the neutron star rather than the bombardment of X-rays on the companion star of the binary. ScoX-1 has also quasi-periodic oscillations (QPOs) which is attributed to the interaction of the magnetic field with the accretion disc whose period of revolution is not regular.

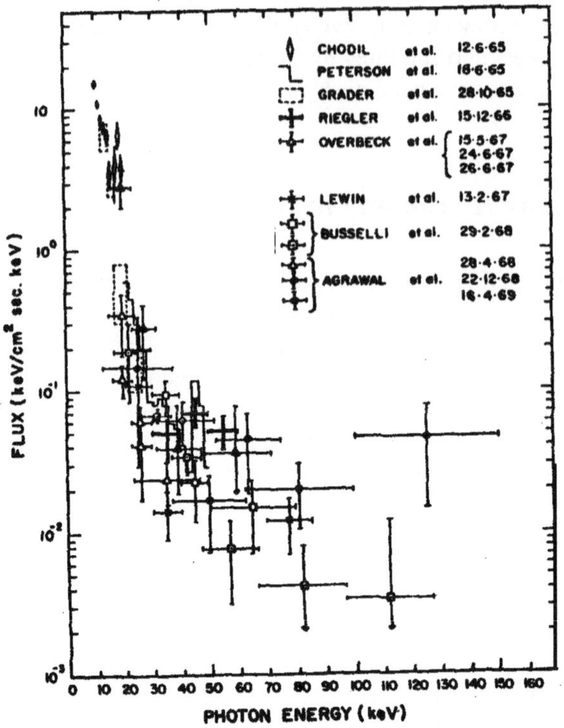

Fig. 4.3 The flattening of the spectrum of ScoX-1 at energies higher than 40 KeV first observed. (Agrawal PC, Biswas S, Gokhale GS, Iyengar TS, Kunte PK, Manchanda RK, Sreekantan BV, Astrophys. Space Sci. vol. 10. pp. 500-507, 1971)

The next source that was observed in the balloon flights from Hyderabad by the TIFR group was CygX-1. One of the most interesting aspect of this source was the recognition of sudden shift in the level of intensity during the course of the flight. Detailed observations with many of the satellite and balloon flights carried out subsequently revealed that CygX-1 is highly variable on time scales of even a fraction of a second. With the identification of the optical counterpart of CygX-1, it became evident that it is a binary system with a 5.6 days period. The determination of the binary parameters enabled mass limits to be set on the two components. A very conservative lower mass limit of the X-ray emitting companion was greater three solar masses. The mass of the non-X-ray emitting companion was very much higher than two solar masses. With these mass limits and the observed time scales of X-ray variations strongly suggested that CygX-1 was most likely a black hole the first candidate for such an object. It was also found to satisfy other criteria for a black hole candidate namely a bimodal spectrum, ultra soft high state, power law tail and flickering in <10 millionth of a second. A complex three state spectra is revealed for CygX-1 up to 1.5 MeV by HEAO-3 data. A bump around 1 MeV is considered as additional support for identifying it as a black hole.

The next source looked at Hyderabad was the Crab Nebula which has been called TauX-1 since it is in the constellation of Taurus. The Chinese royal astronomers had observed according to Chinese astronomical records, the sudden appearance of a new bright star in this constellation on July 4, 1054 AD. This sudden brightness was due to the explosion of a star into a Supernova. The star appeared as bright as Jupiter even during day time for several months. Immediately after the discovery of X-rays from this region, it had been pointed out by Chiu, Shlovsky and Hayakawa independently that the X-ray source could be the star remnant, a neutron star at a temperatures of 10 millions degrees. At the suggestion of Shlovsky, the NRL scientists (Bowyer et al. 1964) made a rocket experiment at the time of lunar occultation of the Nebula on July 7, 1964 and found that the size of the X-ray source in the Nebula is quite big. This ruled out the hypothesis that the source of X-rays in a neutron star. Thus X-ray astronomy lost the opportunity of detecting a neutron star first. It was left to Anthony Hewish to find the pulsar in the Nebula and get the credit for the discovery of neutron star by radio astronomical observations. While the neutron star is also an X-ray source and a pulsar in X-rays, the Nebula itself is emitting X-rays and masked the identify of the central star. The pulsations in X-rays were discovered later.

During 1975, when the Crab Nebula was occulted by the moon again, in an international collaboration between TIFR, the Institute of Aeronautics and Space Sciences, Tokyo, and the University of Nagoya, two hard X-ray telescopes were flown from Hyderabad within a few minutes of each other and the structures of the hard X-ray emitting region of the Crab Nebula were determined (Figure 4.4).

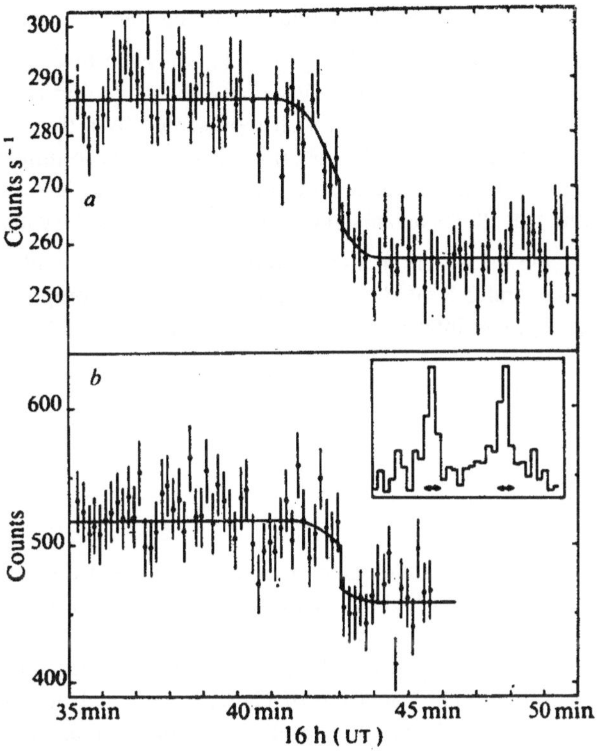

Fig. 4.4 The lunar occultation of the Crab Nebula in hard X-rays observed from a balloon experiment launched from Hyderabad on January 24, 1975 (Fakuda Y, et al. Nature 225:465)

Another X-ray source whose hard X-ray spectrum was first determined by TIFR experiments in 1973 at Hyderabad was Her x-1. Unfortunately a second experiment in 1975 which had a specifically designed detector to look for the line structures at 53 KeV failed because of the collapse of the aluminum Gondola in with the X-ray detector had been mounted on its orientation platform. In 1978, Trumper et al. discovered the 53 KeV cyclotron line from HerX-1. For the first time this observation established unambiguously the existence of a magnetic field as high as 1012 gauss in the neighbourhood of neutron stars.

The UHURU had established in 1972, that HerX-1 has three distinct periodicities (i) a 1.2 seconds pulsation, (ii) a 1.7 days binary orbit period and (iii) a 35 days cycle that modulates the intensity in an irregular period of approximately 9 bright days and 26 dim days. The slowing down of the X-ray pulsar in the HerX-1 is rather erratic.

The pulse profile also is known to fluctuate. Interestingly, HerX-1 has been detected as pulsating in the TeV energy range also. There is also an interesting episode in which HerX-1 showed a burst of 20 minutes duration in the TeV energy range observed by the TIFR group at Pachmarhi (Viswanath et al. 1989). Occasionally even 10^{15} eV emission has been observed from this source (Dingus et al. 1988; Gupta et al. 1971, 1978). An anomaly is that the associated showers contain lot of muons indicating that the showers are unlikely to be due to gamma-rays.

4.4 Soft X-ray Experiments from Thumba Equatorial Rocket Launching Station

The equatorial latitudes of Indian rocket launching stations at Thumba proved particularly advantageous for soft X-ray astronomy experiments because of the low background due to cosmic ray induced secondaries and also electron precipation is a minimum compared to high latitude launching stations. Several large area thin window X-ray proportional counters were launched from TERLS by the TIFR group and about 40% of the sky was mapped in the energy range 0.1–2.5 KeV. These early results on soft X-rays in the energy range 0.1–0.4 KeV showed that the spatial distribution in the north galactic hemisphere has a patchy structure with intensity generally increasing towards higher latitudes. The observations also revealed prominent limb brightening in the north polar spur and a hot spot in Eridanus in the sourthern galactic hemisphere.

In a flight made on January 21, 1973 (21.40 UT), from Thumba the low energy spectra of ScoX-1 and TauX-1 were measured in the energy range 0.5–10 KeV with two banks of thin window proportional counters. The results showed that the spectrum of ScoX-1 is well represented by a thin source bremstrahling with a plasma temperature of $7 \times 10^{7o}K$ attenuated by the absorpion due to a gas column of 2.4×10^{21} H atoms/cm^2. The upper limit to the variability of the X-ray absorption was ±50% at 95% confidence limit. It may be mentioned variability of as much as 200% has also been reported by other investigations.

In the case of Crab Nebula, the measurements showed that the spectrum could be represented by a power law with an exponent $-2:1$ beyond 2 KeV. The absorption of the soft component below 2 KeV is clearly seen in the experiment and leads to an inter-stellar column density of 3.5×10^{21} H atoms in the direction of the Crab (VSI).

4.5 Satellite Experiments

In the very first Indian satellite Aryabhata launched by Intercosmos rocket from the Soviet Union, the Physical Research Laboratory group of Ahmedabad had two X-ray astronomy payloads to cover the energy range 2.5–155 KeV (Rao et al. 1979; Kasturirangan et al. 1976). The energy range 2.5–18.75 KeV was covered by the a proportional counter and the higher energy range by a scintillation counter. Only 10 days observation was possible due to power supply failure. The spectrum of CygX-1 derived from this experiment is shown in Figure 4.5 and compared with the Ariel V satellite results. The spectrum as derived from Aryabhata was a power low with an exporter 0.7 ± 0.2 and estimated intensity of 0.84 ± 0.3 photons/cm^2 sec. Since CygX-1 underwent a major upward transition in intensity in April 1975 just immediately after Aryabhata observation it is interesting to note that CygX-1 exhibited a hard spectrum just before transition and softening other sources like GX17+2 and GX9+9 were also observed in this period.

4.5.1 The First Indigenizes Indian X-ray Astronomy Satellite Experiment (IXAE)

The IXAE was a collaborative effort of TIFR and ISRO Satellite Centre (ISAC) and its principal objective was the study of periodic and aperiodic variability of different types of X-ray binaries. This included measurement of pulsation periods and their rate of change in X-ray pulsars, detection and measurement of QPOs in black hole and neutron star binaries, study of flaring and sporadic variability, detection and study of new X-ray transients, etc. The main component of IXAE instrument was a set of three identical, co-aligned pointed proportional counters (PPCs) with a total effective area of about 1,200 cm^2 and a field of view of $2.3° \times 2.3$ defined by a honey comb-shaped passive collimator. Details of IXAE instrument are given by Agrawal et al. 1976.

The IXAE was launched aboard the IRS-P3 satellite by the Polar Satellite Launch Vehicle (PSLV) on March 21, 1996 into 830 km circular polar orbit, with an inclination of 98° to the equatorial plane. The polar orbit of the satellite produces high and variable background in PPCs, due to high flux of electrons at latitudes $> 45°$. A majority of the orbits also pass through the South Atlantic Anomaly (SAA) region, which is a zone of high fluxes of charged particles. This restricted the useful observation time to typically 20 minutes per orbit and 6 orbits per day.

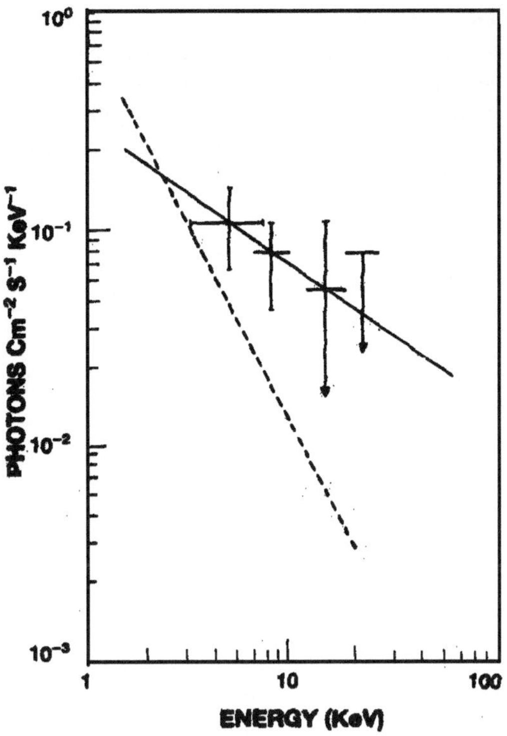

Fig. 4.5 Aryabhata observations on CygX-1 (thick line) before the intensity transition compared with the Aerial V observations (dotted line) after the transition. (K Kasturirangan, SC Chakravarty, MG Chandrasekhar. Science with Space Platforms- Indian Efforts in Retrospect & Prospects, in: Space - In Pursuit of New Horizons. A festschrift for Prof. UR Rao (eds. RK Varma, K Kasturirangan, US Srivastava, BH Subbaraya), Ind Acd Sci pp.155-181)

4.5.1.1 Studies of the Micro-Quasar GRS 1915+105

The X-ray source GRS 1915+105 discovered in 1992 was found to be a variable radio source that exhibited quasar-like superluminal motion and hence, termed as a micro-quasar. Not much was known about its X-ray behaviour till 1995, when Rossi X-ray Timing Explorer (RXTE) and independently IXAE detected rapid erratic X-ray intensity variations and also discovered low frequency QPOs. This object was repeatedly observed with the IXAE over a 5-years period and a host of new and interesting phenomena were detected in it. Chaotic and rapid intensity variations over time scale as small as 100 ms were detected for the first time in this source from IXAE observations made during July 23-27 1996. A strong QPO peak at ~ 0.7 Hz was also detected with fractional root mean square (rms) amplitude of 10% in the power density spectra of all the PPCs. The QPO frequency was found to drift in the range of 0.62–0.82 Hz (Paul et al. 1997). The rapid variability, power density spectrum similar to that of the well known black hole binary CygX-1, high X-ray luminosity near super-Eddington and its peculiar radio features, strongly suggested that the X-ray source in GRS 1915+105 is a black hole.

More detailed follow-up observations of the source were made during June 12–29 1997. It was found to be in a bright state, producing strong quasi-periodic bursts during June 12–17 and June 23–26, with a recurrence time of ~ 45 s with slow rise and fast decay. A sample of these bursts is shown in Figure 4.6, taken from Paul et al. 1998. The slow rise time of the bursts has been explained as arising due to the free-fall time of the mater, ejected from the accretion disk around the black hole and fast decay due to sudden disappearance of the matter behind the event horizon.

A more detailed and thorough analysis of different types of X-ray bursts observed with the IXAE was performed by Yadav et al. (1999) who found a strong correlation between the quiescent interval preceding a burst and the burst duration for the quasi-regular and irregular bursts. It was concluded that the source switched back and forth between the low-hard state and the high-soft state in a very short time, when the accretion rate was at a critical value, resulting in the production of the bursts. Another interesting discovery was the detection of X-ray dips in the light curves of this source in the data of June 6–17 1997, when the source made transition from a low-hard state to a chaotic state. The dips were detected on most of the days and their duration was in the range of 20–160 second. Onset of X-ray dips was followed by the occurrence of a huge radio flare. This led to the inference that the dips are the cause of the mass ejection, due to evacuation of matter from the accretion disk around the black hole and a superposition of a large number of the dips leads to production of radio flare and jet in the source (Naik et al. 2001). Outside the dips ~ 4 Hz QPOs were detected, similar to those seen in the low-hard state, but the QPOs were absent in the dip region.

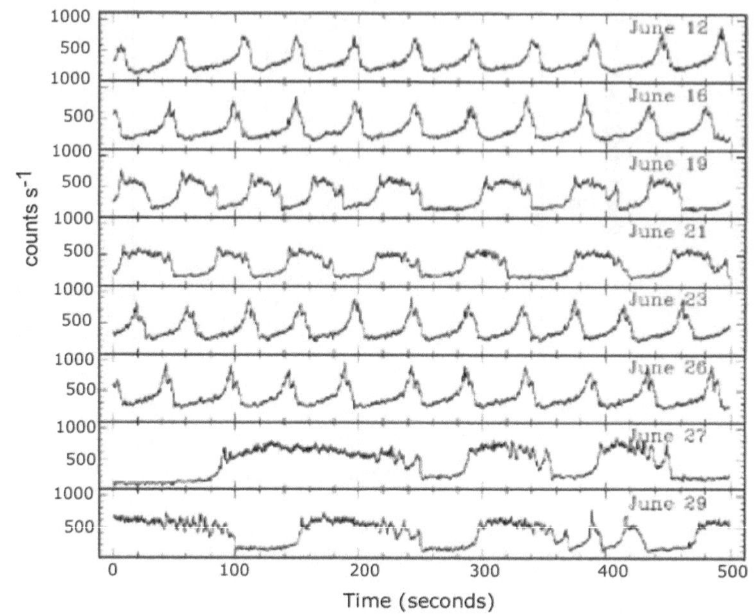

Fig. 4.6 A series of quasi-periodic X-ray bursts, irregular and long bursts detected during June 12-29, 1997 from the micro-quasar GRS 1915+105 by the IXAE on the IRS-P3 satellite, are shown in this figure

4.5.1.2 IXAE Observations of Some Other Black Hole Binaries

During the performance verification phase of IXAE, CygX-1 was observed during May 1-11, 1996 when it was in a low-hard state at a flux level of $\sim 0.3 - 0.5$ Crab. It was again observed in July 5-8, 1996 period, when it had moved to the bright state and was measured to have an intensity of $\sim 0.8 - 1.1$ Crab. The light curves in both the states show chaotic and rapid variability, with flaring activity typical of CygX-1. Bursts of 0.1 second and longer duration were present in both the states, but were more frequent in the hard state (Rao et al. 1998).

CygX-3 was another black hole object studied with the IXAE. Its intensity modulates with the binary period of 4.8 hours. The evolution of the orbital period of CygX-3 was investigated from measurements of the binary period during June 3 – 13, 1999 and again during October 11 – 14, 1999. By combining the measurements obtained from IXAE with those from other earlier X-ray missions, rate of change of orbital period was derived. This leads to a value of 1.05419×10^{-6}/year for the evolutionary time scale of the orbital period. Mass loss from the companion star is inferred to be the most likely cause for the orbital decay of CygX-3.

The IXAE also observed two X-ray transients namely XTE J1748-288 and XTE J2012+381 during their outbursts and monitored their intensities. Transient XTE J1748-288 was observed during its decay phase. Its brightness decayed exponentially with a decay time of 19 ± 1.6 days. There was no indication of any short-term variability and no QPOs were detected from the power density spectrum (PDS) of the source. The second transient XTE J2012+381 underwent outburst on 24 May 1998 as reported by the All Sky Monitor on the RXTE. It was observed with the IXAE during June 2–10, 1998. Its light curve also showed exponential decay of the flux.

4.5.1.3 Studies of X-ray Pulsars with the IXAE

The slow pulsar in the X-ray binary 4U 1907+09 with a pulsation period of 440 second and a binary period of 8.4 days was observed in August 1996 and again in June 1998. X-ray pulsations were clearly detected with a double-peak pulse profile. The primary pulse had an asymmetric shape and was separated from a weaker and broad secondary pulse by a dip. From accurate measurements of the pulsation period for the two observations, the spin-down rate of the neutron star was measured to be 0.23 ± 0.01 sec/year consistent with the earlier measurements, indicating that the pulsar has been monotonically spinning-down, since, its discovery. During August 1996 observation, IXAE detected a flare from 4U 1907+09 at a peak flux level of 88 milliCrab (Mukerjee et al. 2001) shown in Figure 4.7. Analysis of the outburst data revealed transient oscillations with a period of 14.4 seconds, similar to the 18.2 seconds oscillations reported from the RXTE observations during a flare that occurred in February 1996.

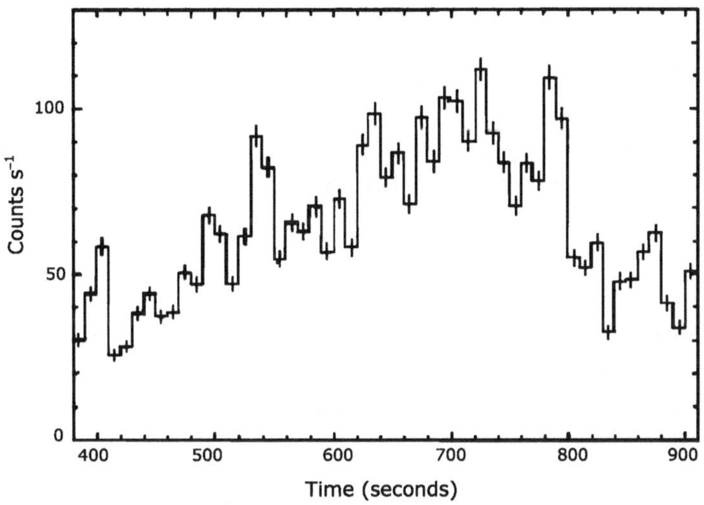

Fig. 4.7 Light curve of the X-ray flare in the X-ray binary 4U 1907+09 detected by the IXAE on IRS-P3 satellite in August 1996

Transient pulsar Cepheus X-4 was observed with the IXAE during the decay phase for 3 days. Pulsations were detected at the known period of 66 seconds and the pulse profile showed luminosity-dependent changes. During the declining phase the main-pulse, dominated double pulse profile changed to inter-pulse dominated profile, when the X-ray luminosity fell to a value of 6×10^{35} ergs s^{-1}.

Another transient X-ray pulsar XTE J1946+274, which is a Be X-ray binary with an orbital period of $\sim$80 days, was also observed with the IXAE in September 1999 and in June to July, 2000. Strong pulsations with 15.8 seconds period were detected in both the observations with a double-peak profile and pulse fraction of $\sim$ 30%. By combining data from the IXAE and those from Burst and Transient Source Experiment (BATSE) and other satellites, the period derivative value was found to be 1.27×10^{-9}s s^{-1} consistent with constant spin-up of the neutron star.

4.5.2 Indian Multi-wavelength Astronomy Satellite 'Astrosat'

Astrosat is conceived to be a multi-wavelength astronomy mission that will cover soft X-rays (0.3-8 KeV), hard X-rays (10-100 KeV), near and far ultraviolet bands (120-300 nm) and visible band. The primary goal of the Astrosat is defined to be multi-wavelength studies to be realised by making simultaneous observations, with a set of co-aligned X-ray and UV instruments. The other main objectives of Astrosat are high time resolution X-ray variability studies, low and medium resolution measurement of continuum and spectral features, medium resolution X-ray imaging studies and simultaneous imaging and photometric observations in the visible, near-UV and far-UV bands for different class of X-ray and UV sources. Astrosat has been designed for studies of time variability phenomena, like pulsations, high frequency QPOs, flaring activity, etc. in X-ray binaries and other variable sources, spectral measurement of all type of X-ray sources and obtaining energy spectra of these sources over five decades, in energy extending from visible to hard X-ray region through simultaneous observations. Astrosat will have improved sensitivity in the hard X-ray band that will facilitate detection and study of cyclotron lines, detection of non-thermal component in the spectra of black hole binaries, supernova remnants, clusters of galaxies, etc. It has been designed to have better sensitivity for detection of QPOs above 10 KeV in X-ray binaries.

Astrosat is a national project, with international contributions, in which, large number of Indian Institutions like TIFR, ISAC, IIA, RRI, IUCAA, etc. are participating and contributing to the development of hardware. There are also two foreign partners, the Canadian Space Agency (CSA) and Leicester University (LU), UK facilitating the development of two important parts of the Astrosat instruments, namely, the photon counting detectors for the ultraviolet imaging telescope (UVIT) by CSA and the X-ray charge coupled device (CCD) camera for the soft X-ray imaging telescope (SXT) by LU. Several other centres of ISRO are designing and fabricating various components and sub-systems of the Astrosat instruments.

4.6 Astrosat Instruments

The instruments planned for launch on board Astrosat are: (i) large area X-ray proportional counters (LAX-PCs); (ii) cadmium-zinc-telluride imager (CZTI); (iii) (SXT); (iv) scanning X-ray sky monitor (SSM); (v) UVIT. A description of the Astrosat instruments and their characteristics is given by Agrawal (2006).

4.6.1 Astrosat Mission

The Astrosat will be a three axis stabilised satellite, with a capability for orientation maneuvers and attitude control, using reaction wheels and magnetic torquers, which get input from three gyros and two star sensors. It will have a pointing accuracy of about 1 arc second. A solid-state recorder with 120 Gb storage capacity will be used for on board storage of data. Two carriers, at a rate of 105 Mb/sec, will transmit the payload data. The total mass of Astrosat observatory is estimated to be 1,600 kg, including 868 kg mass of the scientific instruments. It will be launched in to a 650 km altitude circular orbit, with an orbital inclination of 8° by the well-proven PSLV from Satish Dhawan Space Center (SDSC), Shriharikota by the November 2009. The Astrosat will have a minimum mission life of 5 years.

4.6.2 Science Expected from Astrosat

Multi-wavelength studies will be a unique capability of Astrosat that will improve the understanding of the radiation processes and the environment in the vicinity of the central compact objects in the AGNs. Observations of X-ray binaries will lead to understanding of the nature, environment, site and geometry of X-ray and UV emission of the compact objects. Variability studies over a wide spectral and time domain will probe the nature of the sources and the cause of variability. Detection and detailed studies of kHz QPOs in hard X-rays is an important object of Astrosat that is the key to probe the accretion flows closest to the compact source. One will be able to successfully search kHz QPOs from the X-ray sources with LAXPC, if the source intensity is above 50 milliCrab.

The X-ray spectral measurements of the continuum and lines in 0.5–100 KeV interval, from simultaneous observations will reveal origin of the different components of the spectra and parameters of the radiation processes. With an exposure of 1 day, LAXPCs will provide spectrum with good statistical significance for a 0.1 milliCrab intensity X-ray source. The CZT imager will be able to detect a source of 0.5 milliCrab in 1,000 seconds and obtain a good spectrum in 1 day of observation. The sensitivity of the LAXPCs and the CZT array for measurement of magnetic field of neutron stars will be superior to that of any other existing experiment. A simulation of the expected signal in the LAXPC array for the cyclotron line fluxes detected from the X-ray pulsar 4U0115+63 with RXTE and comparison with the observed spectra from RXTE-PCA and BeppoSAX PDS (Heindel et al. 1999) shows that the cyclotron lines will stand out very clearly in the LAXPC spectrum. Spectroscopy of hot thin collisional plasmas in galaxies, clusters of galaxies, supernova remnants and stellar coronae, and photo-ionized matter in accreting white dwarfs, neutron stars, black holes and AGNs would be carried out with SXT. With an energy resolution that is 10–50 times better than that of the proportional counters, SXT will separate the line emission and absorption components from the continuum in all known varieties of objects.

The imaging UVIT observations with (~ 2) arc sec angular resolution will measure the morphology and energy distribution of galaxies in the local region. It will also study star bursts in distant galaxies and map the ionized gas in them. It will map the galactic H II regions, planetary nebulae and supernova remnants in our galaxy well as well as those in the nearby galaxies in various emission lines, e.g. CII (235 nm), CIII (190.9 nm), CIV (155 nm), OII (247 nm), etc. to map the elemental distribution and the physical condition of the gas. By studying early type hot OB stars in our galaxy and their distribution in nearby galaxies, one will be able to obtain the star formation histories and enrichment of gas.

4.7 Concluding Remarks

Experimental efforts in India in the field of X-ray astronomy began in 1968 within a few years of the discovery of the first X-ray source ScoX-1, with X-ray telescopes carried on stratospheric balloons flown from Hydreabad followed by rockets flown from TERLS and later by satellites. The institutions that have played a major role in the development of X-ray astronomy in India are the TIFR, the Physical Research Laboratory (PRL) and the ISAC. Only a small fraction of the results in the long series of experiments carried out by these groups over the past four decades have been presented just to give a flavour of the type of results obtained and published in journals. The significance of these results have to be judged in the context of the time at which they were published since in later years, the field of X-ray astronomy was flooded by a large number of dedicated X-ray astronomy satellites equipped with instruments of higher sophistication, higher sensitivity and longer exposure times and naturally the results have a better statistical weightage. What is gratifying however is that some of the important trends discovered in the early experiments, to which there has been Indian contributions too, have not been contradicted and in fact have been well substantiated. The most significant trends discovered in the first few decades of X-ray astronomy are: (i) the X-ray sources are a class by themselves and not always objects which have special significance in optical and radio astronomy, (ii) X-ray results brought special focus on transient phenomena and bursting activities in the sources and led to the recognition of compact binary systems in which accretion of matter played an important role in generating high energy phenomena, (iii) the identifications of black holes candidates as one the companions in the binary systems first came from the study of X-ray timing records, (iv) definitive evidence for the presence of magnetic fields of strength as high as 10^{12} gauss around neutron stars came through the registration of the cyclotron lines in X-ray sources and (v) X-ray results focused attention on the need for simultaneous multi-wavelength observations.

The Indian X-ray astronomy satellite (Astrosat) which is to be launched soon has been designed with the state-of-the-art detector systems and electronics and is expected to yield important results in the multispectral bands in the coming years.

Acknowledgements In preparing this chapter I have liberally made use of the material published earlier in my own articles and also those published in association with my colleagues at TIFR, especially with Prof. P.C. Agrawal to whom I am greatly indebted. I would like to thank Prof. T. Padmanabhan for inviting me to write this chapter. Smt. V.B. Mariayammal has provided all the technical help in the preparation of this chapter.

References

1. Agrawal PC, et al. (1971) Astrophys. Space Sci. **10**: 500-507
2. Agrawal PC, et al. (1976) J. Korean Astron. Soc. Supplement, **29**: 429
3. Agrawal PC (2006) Advances in Space Research **38**:2989
4. Bradt H, et al. (1968) Space Sciences Review **8**: 471-506
5. Bowyer CS, et al. (1964) Nature (London) **201**:1307
6. Bowyer S, et al. (1964) Science **146**: 912
7. Burnight TR (1949) Phys. Rev. **76**: 165
8. Dingus BL, et al. (1988) Phys. Rev. Rep. **60**:1785, Phys. Rev. Rep. **61**:1906
9. Gursky H, et al. (1966) Ap. J. **144**:1249

10. Giacconi R, et al. (1962) Phys. Rev. Lett. **9**, 439
11. Gupta SK, et al. (1971, 1978)Astrophys. J. **354**, L13; Astrophys. J. **221**: 268
12. Heindel WA, et al. (1999) Ap. J., **521**:L49
13. Kasturirangan K, et al. (1992) Indian Acd. Sci., pp.155-181
14. Mukerjee K, et al. (2001) Ap. J. **548**:368
15. Naik S, et al. (2001) Ap. J., **546**:1075
16. Oda M, et al. (1967) Ap. J. **148**: L5-L11
17. Paul B, et al. (1997) Astronomy and Astrophysics, **320**:L37
18. Paul B, et al. (1998) Ap. J. **492**:L-63
19. Purcell JD, et al. (1949) Phys. Rev. **76**:165
20. Rao, et al. (1979) Proc. Indian Aca. Sci., Aryabhata Project, pp. 215-227
21. Rao AR. et al. (1998) Astronomy and Astrophysics , **330**:181
22. Trumper J. et al. (1978) Astrophys. J. **219**:L105
23. Viswanth PR. et al. (1989) Ap. J. **342**:489-492
24. Yadav JS, et al. (1999) Ap.J. **517**:935

Chapter 5

Growth and Development of Radio Astronomy in India

Govind Swarup

5.1 Introduction

5.1.1 Early Years of Radio Astronomy

Radio astronomy is the study of the universe through naturally produced radio waves reaching us from a variety of celestial objects. Over the last 60 years, many extraordinary celestial objects and phenomena have been discovered by radio astronomers that have revolutionized our understanding of the universe. This new science was born when Karl Jansky (1933) serendipitously discovered that radio emission in the form of radio noise was being emitted from the direction of the Milky Way, our galaxy. This discovery remained unnoticed by optical astronomers and physicists for many years. In 1935 Grote Reber, a young amateur radio engineer, constructed a 10 m diameter parabolic dish in his backyard and succeeded to make a map of the Milky Way (Reber 1940). Development of radar during World War II resulted in the discovery of radio emission from the sun (Hey 1946a; Reber 1944; Southworth 1945). The publication of these results soon after the end of the war, led to the establishments of active radio astronomy groups mainly in Australia and UK, who investigated the nature of radio emission from the sun and also discovered several discrete radio sources associated with objects in our galaxy and distant galaxies. These remarkable discoveries led to worldwide interest in the great potential of the radio window of the electromagnetic spectrum for exploration of the universe, resulting in development of many outstanding radio telescope facilities in the world as well in India.

In 1945 Australian radio astronomers found that radio emission from the sun had two main components: (a) the radio emission from the quiet sun was estimated to have a temperature of $\sim 10^6 K$ and (b) that from active sun (solar radio bursts) exceeded $10^{11} K$. Ginzburg (1946) and Martyn (1946) independantly considered thermal emission to arise from the solar corona with kinetic temperature of $\sim 10^6 K$. Saha (1946) considered excitation of the energy levels of the nuclei and atoms and molecules by a strong field near sunspots but its contribution was not considered significant by observers (Pawsey 1950). Saha (1946) also postulated radiation at the gyro frequency but its escape was not permitted by severe attenuation at the plasma frequency layer at higher levels (Martyn 1947).

The discoveries of radio emission from celestial bodies were keenly noted by astronomers, physicists and students in India. M. K. Das-Gupta from the Institute of Radiophysics and Electronics (IRPE), Calcutta, went to work at the Jodrell Bank Observatory in UK and made a path-breaking discovery that the strong radio galaxy, Cygnus A, was a double radio source (Jennison and Das-Gupta 1953). M.R. Kundu after graduating from IRPE went to France and got the Ph.D. degree working on a high resolution radio interferometer operating at centimetre wavelengths. T.K. Menon went to Harvard in 1951.

National Centre for Radio Astrophysics, Tata Institute of Fundamental Research, Pune University Campus, Pune–411 007.
e-mail: gswarup29@gmail.com

5.1.2 Growth of Radio Astronomy in India

Monitoring of the solar radio emission was started at the Kodaikanal Observatory in 1952. In 1952, Sir K. S. Krishnan, Director of the National Physical Laboratory (NPL), New Delhi, attended the General Assembly of the International Radio Scientific Union (URSI) held at Sydney. He was very impressed by the discoveries being made by Australian scientists in the new field of radio astronomy. Krishnan arranged deputation of the author (G. Swarup) of NPL and R. Parthasarathy of the Kodaikanal Observatory for 2 years during 1953–55 under a Colombo Plan fellowship at the Radiophysics Division of the Commonwealth Scientific and Industrial Research (CSIRO), Australia. In 1955, CSIRO agreed to gift to NPL 32 parabolic dishes of 1.8 m diameter, which formed the grating array built by W.N. Christiansen in 1952 at Potts Hill, near Sydney, and were being dismantled as a new radio telescope was being built by Christiansen (Swarup 2006, 2008). On return from Australia, Swarup and Parthasarathy joined NPL in 1955. Since the transfer of the dishes got delayed, they went to USA in 1956. During 1956–58, NPL attracted N.V.G. Sarma and M. N. Joshi after their M.Sc., T. Krishnan after his M.S. from the University of Cambridge and M.R. Kundu after his Ph.D. in France. Later Kundu joined the University of Michigan, Krishnan went to work at the University of Sydney in Australia. Sarma went on deputation for 2 years to work on a radio telescope at the University of Leiden and Joshi to France for a Ph.D. degree in radio astronomy; both returned to NPL in 1962 or so. Later during 1963–1965, Swarup, Sarma, Joshi and Kundu joined the new radio astronomy group that was formed at the Tata Institute of Fundamental Research (TIFR), Mumbai, in 1963 (see Section 5.2). It may be noted that although NPL was not able to provide suitable support for the growth of radio astronomy therein, it nurtured a group that later joined TIFR. Over the last 45 years, two world class radio astronomy facilities have been built in India by TIFR at Udhagamandalam (Ooty) and near Pune as described in Section 5.2. Astronomy with the Glant Metrewave Radio Telescope (GMRT) is described in Section 5.3. The development of radio astronomy at the Indian Institute of Astrophysics (IIA), Bengaluru, is described in Section 5.4, at the Raman Research Institute (RRI), Bengaluru, in Section 5.5, and at the Physical Research Laboratory (PRL), Ahmedabad, in Section 5.6. Conclusions and future thrusts are discussed in Section 5.7.

5.2 National Centre for Radio Astrophysics of TIFR (NCRA/TIFR)

5.2.1 Brief History

In September 1961, four radio astronomers working abroad, M.R. Kundu, T. Krishnan, T.K. Menon and the author (G. Swarup) wrote a ‘proposal for the formation of a radio astronomy group in India’ and sent it to several scientific organizations in India. Dr. Homi Bhabha, the great visionary and founder Director of TIFR met T.K. Menon in Washington DC in USA and “indicated outlays of Rs. 50–100 lakhs if the group fulfils the expectations”. He sent a telegram to the above four persons on January 20, 1962 stating that TIFR has decided to establish radio astronomy group, a prompt action indeed by any standards in the world (Swarup 1991).

The radio astronomy group got established after I joined TIFR in April 1963. Sarma and Joshi joined in 1964 and played a critical role in the development of the Ooty Radio Telescope (ORT). M.R. Kundu joined the group in 1965 and returned to USA in 1968; T.K. Menon joined in 1971, after $\sim$ 20 years in USA, and returned to join the University of Vancouver in Canada in 1974; contributions by both were very important in the growth of the radio astronomy of TIFR in its early years. In Section 5.2.2 are described four radio telescopes built by the group over the last four decades. Subsequent sections present some of the highlights of the group’s work, which spans a variety of topics in the fields of solar, galactic and extragalactic radio astronomy. Radio Astronomy Group was originally located at TIFR, Bombay (now Mumbai). The Radio Astronomy Centre (RAC) of TIFR was formed at Ootacamund (now Ooty) in 1966. The TIFR Centre at Bangalore for radio astronomy was established at Bangalore in 1977 and got shifted to the National Centre for Radio Astrophysics of TIFR at Pune that was formed in 1987. NCRA has now academic headquarters of the group at Pune with observational facilities at Ooty in Tamil Nadu and the GMRT centre $\sim$ 80 km north of Pune near Khodad in Maharashtra.

5.2.2 Research Facilities

5.2.2.1 The Kalyan Radio Telescope

As a first step, the newly formed radio astronomy group of the TIFR, set up a grating-type radio interferometer at Kalyan near Bombay in 1965 for observing the sun at a frequency of 610 MHz (Swarup et al. 1966). The interferometer consisted of 32 parabolic dishes of 1.8 m diameter that were gifted by CSIRO, Australia to the NPL and later transferred to TIFR. Twenty-four dishes were placed along a 630 m east-west baseline and eight along a 256 m north-south baseline, giving an angular resolution of 2.3 arc-min × 5.2 arc-min (Figure 5.1). The telescope was used for studying the quiet and active regions of the sun during 1965–68. It was found that the quiet sun had considerable limb-brightening at 610 MHz and that the solar corona had a temperature of $\sim 10^6$K (Figure 5.2) (Sinha and Swarup 1967). Solar radio bursts were also observed at 610 MHz. The Kalyan Radio Telescope was disbanded in 1968 as the group got involved in the ambitious ORT described in the next Section.

Fig. 5.1 Twenty-four dishes of the EW array of the Kalyan Radio Telescope at 610 MHz (Swarup et al. 1966)

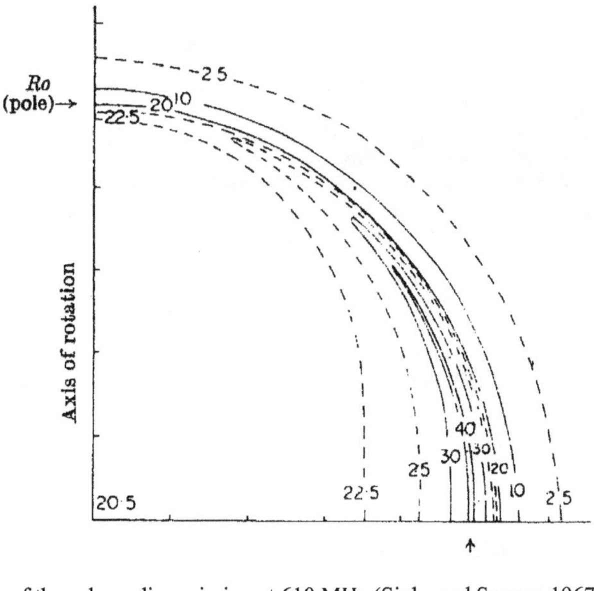

Fig. 5.2 Brightness distribution of the solar radio emission at 610 MHz (Sinha and Swarup 1967)

5.2.2.2 The Ooty Radio Telescope

As stated in the 1961 proposal that was sent to Bhabha and others (see Section 5.2.1), our initial idea for a major radio astronomy facility was to set up a large Mills Cross operating at a wavelength of about $2m$ in India. However, the ionospheric scintillations that are prominent in India due to its close location to the magnetic equator seemed a cause of worry. In early 1960s there was a great controversy between the Big Bang model and the Steady State model for explaining the origin and evolution of the universe. During late 1950s, Martin Ryle and colleagues at the University of Cambridge had plotted number versus intensity of ~ 200 radio sources discovered by them and found an excess of weaker radio sources. By assuming that the weaker radio sources are located far away, Ryle supported the Big Bang model. However, Fred Hoyle, proponent of the Steady State model, questioned the above assumptions. For distinguishing between the above models, measurement of angular sizes of a large number of weak radio sources seemed important but suitable radio interferometers providing arc-sec resolution were not available anywhere in the world at that time. Hence I proposed to exploit the lunar occultation method to measure angular structure of a large number of radio sources that required a large steerable radio telescope. This proposal resulted in the ORT. It was the first major facility in India which firmly established the country on the world map of radio astronomy.

The ORT consists of a 530 m long and 30 m wide parabolic cylinder (Figure 5.3) Its design makes full use of India's proximity to the geographical equator. The unique feature of this telescope is that its long axis is aligned in the north-south direction along a hill which has a natural slope of about $11°$, the geographical latitude of Ooty. Thus the long axis of rotation of the telescope becomes parallel to the earth's rotation axis, enabling the telescope to track radio sources in the sky by a simple mechanical rotation of the parabolic frames in the east-west direction. The pointing in the north-south direction is achieved by electronic phasing of the 1056 dipoles placed along the 530 m long focal line of the parabolic reflector. A useful declination range of $\pm 60°$ can thus be covered. The signals picked up by the dipoles are combined with suitable phase shifters to form 12 independent beams designed to cover the lunar disk. The reflecting surface is made up of $1,100$ thin stainless steel wires running parallel to each other for the entire length of the cylinder. The surface is supported by 24 parabolic frames on towers located 23 m apart. ORT operates at a frequency of 325 MHz (Swarup et al. 1971). It provides an effective collecting area of $\sim 8,500$ m^2. The receiver system was upgraded in 1993 by placing a RF amplifier and a phase shifter after every dipole, increasing its sensitivity by a factor of 4 (Selvanayagam et al. 1993). Recently a digital system has been installed by the RRI group (Prabu 2008). The RF bandwidth of the ORT is about 16 MHz.

Fig. 5.3 ORT consists of a 530 m long and 30 m wide parabolic cylinder with its long axis parallel to that of the earth; the long white streak on right side is reflection of 1,100 stainless steel wires by the sunlight (Swarup et al. 1971)

5.2.2.3 The Ooty Synthesis Radio Telescope

The resolution of a telescope, in units of radian, is given by the ratio of the wavelength of radiation divided by the size of its aperture. At a wavelength of ~ 1 m, we require the aperture size to be nearly 200 km to get a resolution of 1 arc-sec, that is available in fact for a mere 10 cm size of a lens of an optical telescope.

Alternatively, in order to achieve high resolutions at radio wavelengths, radio astronomers have developed radio interferometer systems over the last 60 years, starting from a simple 2-antenna interferometer to the complex synthesis radio telescopes built today. A synthesis radio telescope consists of several antennas located over a large area that measure the coherence pattern of the incoming wave front received from a distant celestial radio source. It is readily shown that a 2-antenna interferometer measures one Fourier component of the brightness distribution of the celestial source with its value depending on the separation of the antennas. With n antennas, we measure $n(n-1)/2$ Fourier components. With the rotation of the earth, the projected separations of the antennas change continuously and thus for an array of only 20 or 30 antennas millions of values are recorded. An inverse Fourier transform is carried out in powerful computers using sophisticated self-calibration algorithms that yields radio image over the entire field of view of the component antennas of the array. By 1980s, several powerful synthesis radio telescopes had been built in UK, Netherlands, and Australia. These were designed to work at cm and dcm wavelengths (at frequencies of $\sim 1{,}400$ MHz to over $10{,}000$ MHz), particularly to obtain high angular resolution for finding fine structure of celestial radio sources. Soon after the completion of the ORT it was decided to build the Ooty Synthesis Radio Telescope (OSRT) operating at 325 MHz (see next paragraph) and later the powerful GMRT operating in the frequency range of $\sim$ 130 MHz to $1{,}430$ MHz (see Section 5.2.2.5), thus complementing radio telescopes elsewhere in the world.

The OSRT used the principle that the effective area of a pair of antennas, A1 and A2 is given by $(2A_1A_2)^{0.5}$. Since ORT has a large area (500 m $\times$ 30 m), we built seven smaller antennas of size 23 m $\times$ 9 m and one of 92 m $\times$ 9 m in an array of $\sim$ 4 km in extent (Swarup 1984). Signals from three nearby antennas were brought to a central receiver system using coaxial cables and that from far away antennas by means of radio-telemetry. OSRT was equivalent to a 4 km radio telescope giving a resolution of about 1 arc-min at 92 cm wavelength. OSRT was used for studying many galactic and extragalactic radio sources. A few of the results obtained are described in Section 5.2.2.3. The OSRT project allowed the group to master complex UHF electronics and image processing techniques. It was dismantled in 1986 in order for the group to concentrate on a much more challenging GMRT project.

5.2.2.4 Giant Equatorial Radio Telescope (GERT)

The success of the ORT led to the idea of building a much larger radio telescope quiet close to the earth's equator, say in Kenya or Indonesia, as a collaborative project between developing countries. The proposal was discussed with A. M. M'Bow, Director-General of UNESCO during his visit to TIFR in 1977. He was very enthusiastic about it and approved a grant of US$ 14,000 for holding an international workshop and visits to a few countries in Africa. Designs were developed and cost estimates were made for a steerable parabolic cylinder of 2 km long and 50 m wide and also 10 smaller cylinders of 100 $m \times 50$ m to be placed up to 14 km away in order to form a synthesis radio telescope. A detailed report was prepared proposing the establishment of an International Institute of Space Sciences and Electronics (INISSE) and the Giant Equatorial Radio Telescope (GERT) by Swarup from India, Odhiambo from Kenya and Okoye from Nigeria (Swarup et al. 1979, 1984; Swarup 1981). During negotiations, Kenya's interest dwindled after President Kenyatta expired in 1978. In 1981 Dr. Hidayat of the Boschha Observatory, Indonesia, arranged a visit by an Indian team to W. Sumatra, where a site was located close to the earth's equator. In August 1983, President Suharto approved half of the estimated cost of US$ 20 million and India was to meet the other half. However, with the success of the VLA in 1980, development of the method of self-calibration and the newly developed optical fibre transmission technique for RF signals, it seemed to us that it should be possible to build a more flexible and powerful synthesis radio telescope at metre wavelengths in India. Therefore, the GERT became the GMRT! The Noble Laureate Tony Hewish, UNESCO Consultant, for GERT wrote to me in 1983 that his enthusiasm for the GERT decreased after Kenya lost interest.

5.2.2.5 The Giant Metrewave Radio Telescope (GMRT)

Although most of the pioneering work in radio astronomy during 1940s and 1950s was done at metre wavelengths, many powerful radio telescopes were built subsequently at shorter wavelengths for obtaining higher angular resolution. *However, there are many outstanding astrophysical problems that can be studied only at metre wavelengths* ($\sim$0.2–10 m), such as observations of the redshifted $1{,}420$ MHz line emission of the neutral hydrogen (HI) from distant celestial objects. Many other phenomena can also be advantageously investigated at long wavelengths, such as studies of Pulsars or diffuse features in the radio brightness distribution of celestial radio sources. These considerations led to the GMRT proposal.

The GMRT consists of 30 fully steerable parabolic dishes, each of 45 m in diameter (Swarup et al. 1991). A noteworthy feature of the GMRT is its hybrid design. Fourteen antennas are located randomly in a compact array of ~1 km × 1 km in size that allows mapping of the extended features of celestial radio sources with arc-min resolutions. Other 16 antennas are located along three Y-shaped arms for providing higher angular resolution (Figure 5.4).

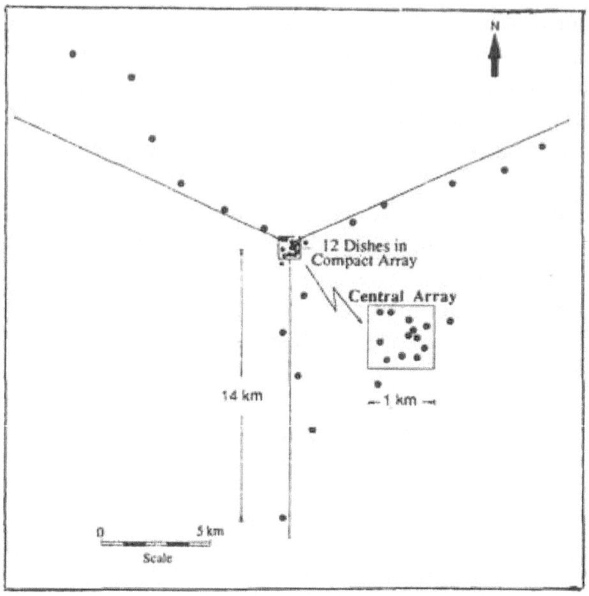

Fig. 5.4 Array configuration of the 30 numbers of 45 m parabolic dish antennas of the GMRT (Swarup et al. 1991)

For parabolic dish antennas operating at microwave frequencies it is required to use aluminium sheets in order to get good reflectivity, but that are subject to large wind loads. For operation at decimetre and metre wavelengths, the reflecting surface of the dishes is often made of wire mesh panels in order to reduce wind loads than that for solid panels yet providing sufficient reflectivity. However, the supporting structure of wire mesh panels is often subject to large wind drag. A design breakthrough was made for the construction of the 45 m dishes. In order to obtain the required curved surface of parabolic dishes using wire meshes of a low solidity, specially developed for the GMRT, an innovative SMART concept (Stretched Mesh Attached to Rope Trusses) was used that also minimized the back up structure of the dishes (Figure 5.5). The wire rope network makes a mosaic of plane facets approximating a parabolic surface. The reflecting surface consists of stainless steel wire mesh of 0.55 mm diameter with a grid size of 10 mm × 10 mm in the inner one-third area, 15 mm × 15 mm in the middle part and 20 mm × 20 mm in the outer part of each of the 45 m dish. Thus it became possible to build 45 m dishes quite economically.

The primary antenna feeds of the GMRT are placed on a rotating turret near the focal point of each dish. The voltage signals received by the dual polarized feeds are amplified by low noise amplifiers and brought to a central point at intermediate frequencies and are applied to a digital correlator system that provides 256 frequency channels over a 32 MHz band. The correlator takes cross products of voltages received from all the 30 antennas, for each of the 256 channels that are finally stored in a computer system for image processing. The GMRT is currently operating in five frequency bands near 150 MHz, 235 MHz, 325 MHz, 610 MHz and 1000–1430 MHz . GMRT is now being upgraded to provide more or less continuous frequency coverage from ~ 40 MHz to 1430 MHz and a bandwidth of ~ 256 MHz or even ~ 400 MHz The effective area of the GMRT antennas is 30,000 m^2 for the lower four frequency bands and ~ 18,000 m^2 near 1,430 MHz. GMRT is the most powerful radio telescope operating at metre wavelengths in the world.

Fig. 5.5 A close view of a 45 m parabolic dish antenna of the GMRT. Ten far away antennas are seen in the background

5.2.3 Major Scientific Contributions by NCRA-TIFR

The research contributions made by the radio astronomy group of TIFR during the last 45 years have been published in over 800 refereed papers. In Section 5.2.3.1 is presented contributions made during the years 1970 to 1998 mostly using the ORT, prior to the operation of the GMRT. In Section 5.2.3.2 are given interplanetary scintillation studies made with the ORT from 1970–2008. Research contributions made during the last 10 years using the GMRT are described in Section 5.2.3.3.

5.2.3.1 Highlights of Astronomy at Ooty

Although the ORT was conceived in 1963 for cosmological investigations (see Section 5.2.2.2), the radio astronomy group of TIFR has made many valuable contributions concerning cosmology, extra-galactic radio sources, galactic radio sources and interplanetary medium (see Sections 5.2.3.2–5.2.3.4). The contributions made during 1963–1988 are given in ~ 250 papers and have been summarised by Swarup et al. (1991) During 1988–1998 there were about 200 papers and nearly 350 during 1999–2008 after the GMRT became operational. Here we give only a few highlights.

5.2.3.2 Radio Astronomy and Cosmology

In 1929, Hubble made a remarkable discovery that further a galaxy is located from us, faster it is moving away. This led to the Big Bang model for the origin and evolution of the universe. In 1948, Hoyle, Bondi and Gold proposed an alternate model, the Steady State model, in which the matter was created not just in the beginning but everywhere, say near the nuclei of galaxies, to explain the expanding universe.

Soon after World War II, a strong radio source was discovered towards the constellation Cygnus (Reber 1944), named later as Cygnus A. The observed flux density of radio waves from the Cygnus A nearly equaled that of the quiet sun, in contrast to the sun being millions of time brighter than stars or galaxies at optical wavelengths (Hey et al. 1946b). Based on its accurate position measured by Smith (1951), Baade and Minkowski (1954) identified it with a faint optical galaxy with a redshift of 0.06 that was several times more distant than that of any known galaxy at that time! The source was shown to be a double radio source with a separation between the two components of ~ 1 arc-min (Jennison and Das Gupta 1953). By 1960 nearly 200 radio galaxies were catalogued using radio interferometers by English and Australian radio astronomers but only ~ 30 could be identified with visible galaxies. It became clear that radio galaxies are located at cosmological distances and therefore their statistical properties could be used for distinguishing between cosmological models. As noted in Section 5.2.2.2, Martin Ryle from Cambridge plotted number counts versus flux density of ~ 200 radio galaxies and concluded support for the Big Bang model. ORT was conceived to measure accurate positions and angular sizes of a large sample of radio galaxies as described below. These observations allowed us to study cosmological evolution of radio galaxies and other active galaxies.

Determination of the redshift of a rather compact quasi-stellar radio source (quasar) 3C 273 as 0.158 by Martin Schmidt (1963) indicated existence of a new class of celestial object in the universe. 3C 273 is optically bright with a magnitude of 13. By $\sim$ 1980s QSOs were identified up to redshift, z $\sim$ 4 and today beyond $z \sim 6$. Many of these sources are radio loud quasars (RLQ) but large numbers are radio quiet quasars (RQQ). High resolution observations made at metre wavelengths by the TIFR group have provided independent support to the unified models concerning radio galaxies and QSOs, as discussed below.

Lunar Occultation Observations: The ORT tracked the moon daily during 1970–1978, (except during 1972–73), for a general survey of the sky that provided one-dimensional brightness distribution along two or more directions across $\sim$ 1,000 galactic and extragalactic radio sources of flux density >0.6 Jy. These observations provided high resolution of $\sim$ 1–10 arc-sec at 325 MHz. It was the best resolution achieved at that time for a large number of weak radio sources. Subrahmanya (1975) developed an innovative 'positivity constraint' for image restoration. The data was used for studies of galactic and extragalactic radio sources and cosmology.

Cosmological Studies: Using the occultation observations, Swarup (1975) derived a relation between angular size, θ, and flux density, S, for radio galaxies for the first time. It was found that the median value of angular size, θm, varies from $\sim$ 100 arc-sec for the nearby strong radio galaxies with $S_m \sim 10$ Jy to 10 arc-sec for $S_m \sim 0.6$ Jy (Figure 5.6). Kapahi (1975) combined the radio source counts, N(S) and $\theta_m - S_m$ relation and concluded that not only the space density of radio sources was higher but also their angular size was smaller at earlier cosmic epochs. This result was not consistent with the Steady State model and indicated evolution of radio sources with cosmic epoch. It provided support to the Big Bang model, independently to the discovery of the cosmic microwave background radiation by Penzias and Wilson (1965). Subsequently $\theta_m - S_m$ relation has been investigated in detail by a number of workers using radio interferometers (Figure 5.7) (see Kapahi 1989 for a review). It was extended up to lower flux densities by Mark Oort with a median value of $\sim$ 1 arc-sec at 1 mJy.

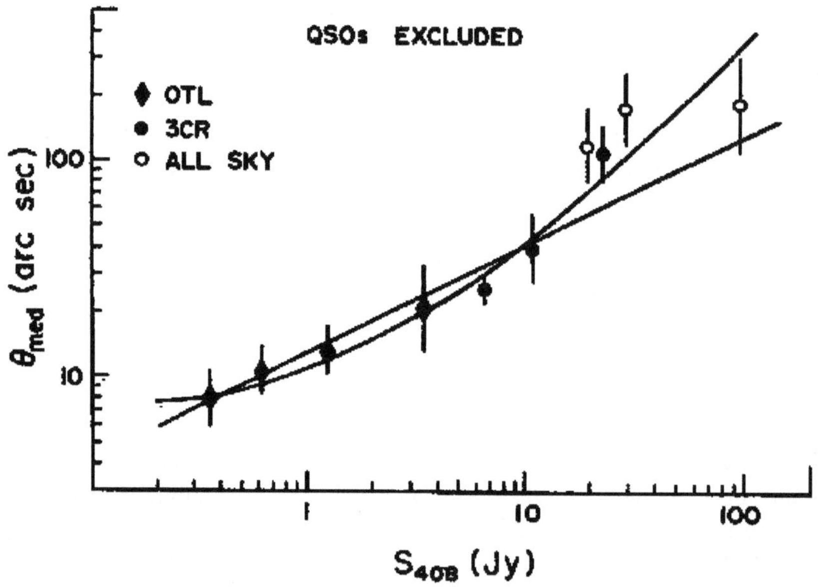

Fig. 5.6 The curved line in the Figure shows a relation between the median value of angular size of radio galaxies versus their flux density (Swarup 1975)

Angular Size-Redshift Relation: Later when redshift estimates became available for a large sample of radio galaxies, it became possible to investigate the angular size-redshift relation directly (Figure 5.8). A steep evolution of linear sizes, l, with redshift z was determined as $l \; \alpha \; (1+z)^{-3}$, and also a mild dependence on radio luminosity, $l \; \alpha \; P^{0.3}$ (Kapahi 1989). During 1976–1988 cosmological evolution of linear sizes and radio luminosity functions (RLF) were investigated in several papers indicating that the linear sizes were smaller on an average at higher redshifts (at earlier cosmic epochs), possibly due to higher density of the intergalactic medium at earlier cosmic epoch.

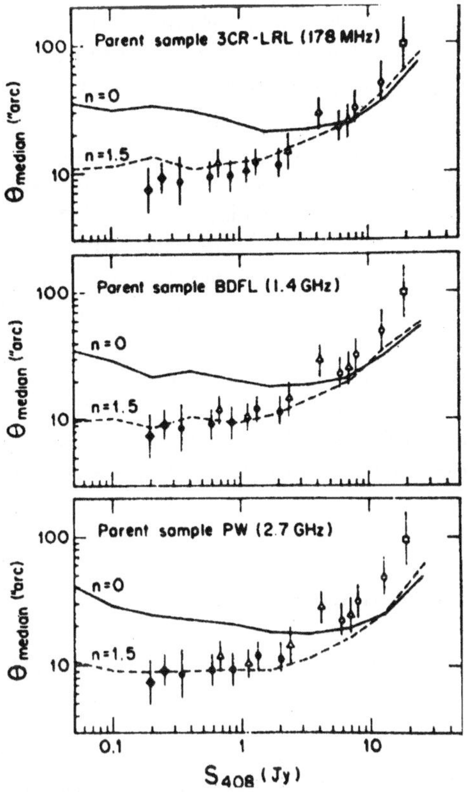

G

Fig. 5.7 Shows a comparison between predictions of evolutionary models of radio luminosity function and the observed angular size-flux density relation (Kapahi et al. 1987)

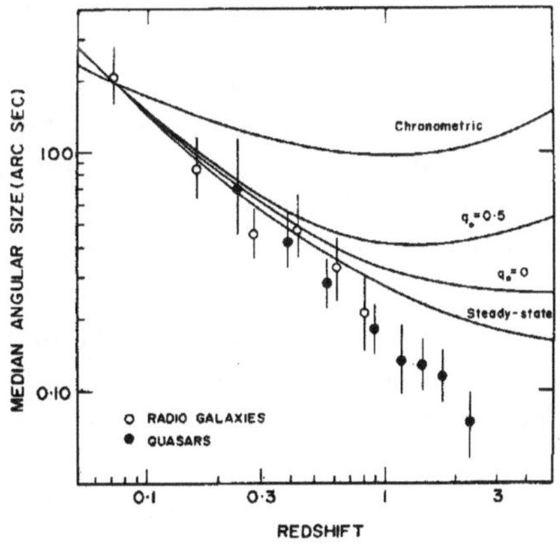

Fig. 5.8 Open and filled circles show values of the observed median angular size of radio galaxies revealing a steep evolution of linear sizes with redshift z (Kapahi 1987)

Spectral Index-Flux Density (α-S) Relation: The flux density values of the Ooty occultation sources at 325 MHz were combined with measurements made at higher frequencies using other radio telescopes (Gopal-Krishna and Steppe 1982). This provided an additional constraint on the RLF. Miley and others have shown that radio sources with steep spectral index, say $\alpha >\sim 1.3$, $(S \propto \nu^{\alpha})$ are often located at higher redshifts, allowing to search for radio galaxies at earlier cosmic epochs and thus investigations of cosmic evolution of radio galaxies. Interesting question and possible answer for the dependence of the steeper spectral index for high redshift radio galaxies has been investigated by Kapahi and Kulkarni (1986), Gopal-Krishna (1988), and Athreya and Kapahi (1998).

Search for Primordial Hydrogen: According to the Big Bang model, electrons, protons neutrons and sea of photons formed in the first few minutes after the origin of the universe at temperatures of tens of million degrees and thus constituted $\sim$ 74% of hydrogen $\sim$ 26% of helium and only a trace of other light elements such as lithium and deuterium. After a few million years the universe cooled and electrons and protons combined to form neutral hydrogen. Subsequently structure formed in the universe by gravitational collapse. Thus condensates of massive primordial neutral hydrogen (HI) are expected to occur in the universe prior to the formation of galaxies. Observations were made with the ORT at 325 MHz during 1980s and indicated that Zeldovich Pancake with mass $5 \times 10^{15} M_0$ did not form at $z = 3.3$, as predicted by the hot dark matter models (Subrahmanyan and Swarup 1990). Observations made using the VLA by Subrahmanyan and Anantharamiah (1990) put upper limits to the HI mass of $\sim 10^{14} M_0$. The above studies and also stronger support to the Cold Dark Matter model by astrophysicists by mid 1980s indicated to us the requirement of a much larger radio telescope for exploring the era of galaxy formation and that led to the GMRT project described in Section 5.2.2.5.

The Deuterium Abundance: Deuterium is predicted to have been produced in the first few minutes after the Big Bang. The measurement of its abundance relative to hydrogen is of considerable cosmological importance. Observations by Sarma and Mohanty (1978) from the Ooty group and Anantharamiah and Radhakrishnan (1979) from RRI put an upper limit to the D/H ratio, in contrast to the positive detection reported previously by the Caltech group.

5.2.3.3 Radio Galaxies, Quasars and Active Galaxies

Radio galaxies and quasars are the most energetic celestial objects in the universe. The radio luminosity of the normal galaxies is 10^{37} erg s^{-1}, star-burst galaxies typically radiate 10^{40-41} erg s^{-1} whereas that of radio galaxies and quasars is several magnitude higher up to $\sim 10^{45}$ erg s^{-1}. It is now well established that these highly energetic objects are associated with active galactic nuclei (AGNs) that are powered by massive black holes with masses of tens to hundreds of million solar masses. AGNs give rise to jets of relativistic particles that interact with the intergalactic gas and emit radio waves by synchrotron process (see e.g. Burke and Graham-Smith 1997). In recent years extensive observations have also been done of the AGNs and associated active galaxies at infrared, optical, ultraviolet and X-ray wavelengths. Observations at metre wavelengths have provided valuable complimentary information about these objects such as their age, cosmological evolution, etc.

While the occultation survey discovered mostly uncatalogued radio sources providing their accurate positions and structures, predicted occultation of several well known radio sources were used to compare the brightness distributions at 92 cm wavelength with that available from the aperture-synthesis observations made at shorter wavelengths (Gopal-Krishna and Swarup 1977; Gopal-Krishna 1977). It was found that diffuse features believed to be filled with energy-depleted features did not show the anticipated brightening at metre wavelengths, constraining the evolutionary model of the radio sources (Figure 5.9). However, many radio galaxies do show spatial variation of spectral index that provide estimates of the ages of the radio galaxies. Menon (1976) compared brightness distribution of a selected sample of sources from the occultation survey at 327 MHz with the observations made with the 3-element Green Bank array and derived spectra of various sub-components.

Compact Radio Sources and Gigahertz Peak Radio Sources: A new class of '*compact radio sources*' were identified by Kapahi (1981) from high resolution observations of a complete sample from a 5 GHz survey, made using the Westerbork Array. The much lower incidence of such sources in low-frequency surveys was attributed to a turnover in their spectrum at several hundred MHz. Gopal-Krishna et al. (1983) identified a large number of related and important class of '*gigahertz peaked spectrum (GPS) sources*'. Extensive work has been done on these interesting classes of radio sources by many workers during the last 25 years as they provide valuable information about the intergalactic medium and central regions of young active galaxies.

Relativistic Beaming and Unification of Quasars and Radio Galaxies: Some of the earliest tests to verify the possibility that the radio galaxies and quasars are related were done by Gopal-Krishna et al. (1980), Kapahi (1981b) and Kapahi and Saikia (1982). While radio galaxies are mostly extended double radio sources with relatively weak central component, quasars have prominent central components with prominent jets. Investigations showed that their differences arise due to the viewing angle with respect to the axis of their radio jets consistent with the predictions of the relativistic beaming model and the 'unified scheme' developed independently by Orr and Browne (1982). Observed asymmetries in double radio sources were investigated (Gopal-Krishna 1980; Saikia 1984).

Gravitational Lens 1830–21: This uniquely bright pair of flat-spectrum radio knots separated by just 1 arcsec was discovered serendipitously in the course of a galactic Plane Survey of scintillating extragalactic radio

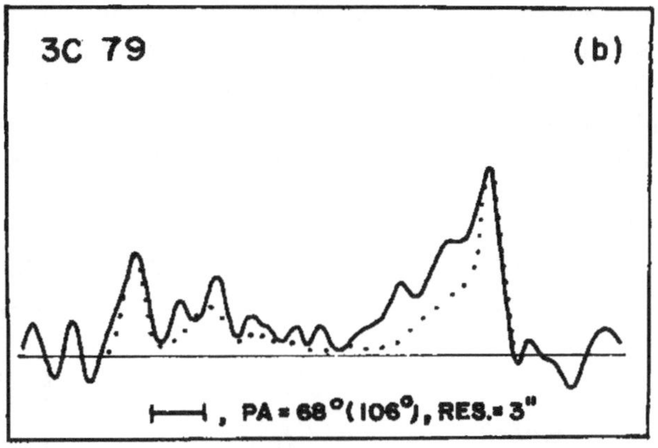

Fig. 5.9 The solid curve shows brightness distribution derived from the Ooty occultation observations at 325 MHz; the dotted curve that derived from strip integration of the 5 GHz Cambridge map (Gopal-Krishna and Swarup 1977)

sources using the ORT (Rao and Ananthakrishnan 1984; Rao and Subrahmanyan 1988). Subsequently, the source was observed with the VLA and was explained by the core of a distant radio quasar being lensed by an intervening galaxy (Figure 5.10) (Subrahmanyan et al. 1990). Jauncey investigated it further and considered it as the first example of Einstein's ring. Extensive studies have been made of this source by many workers.

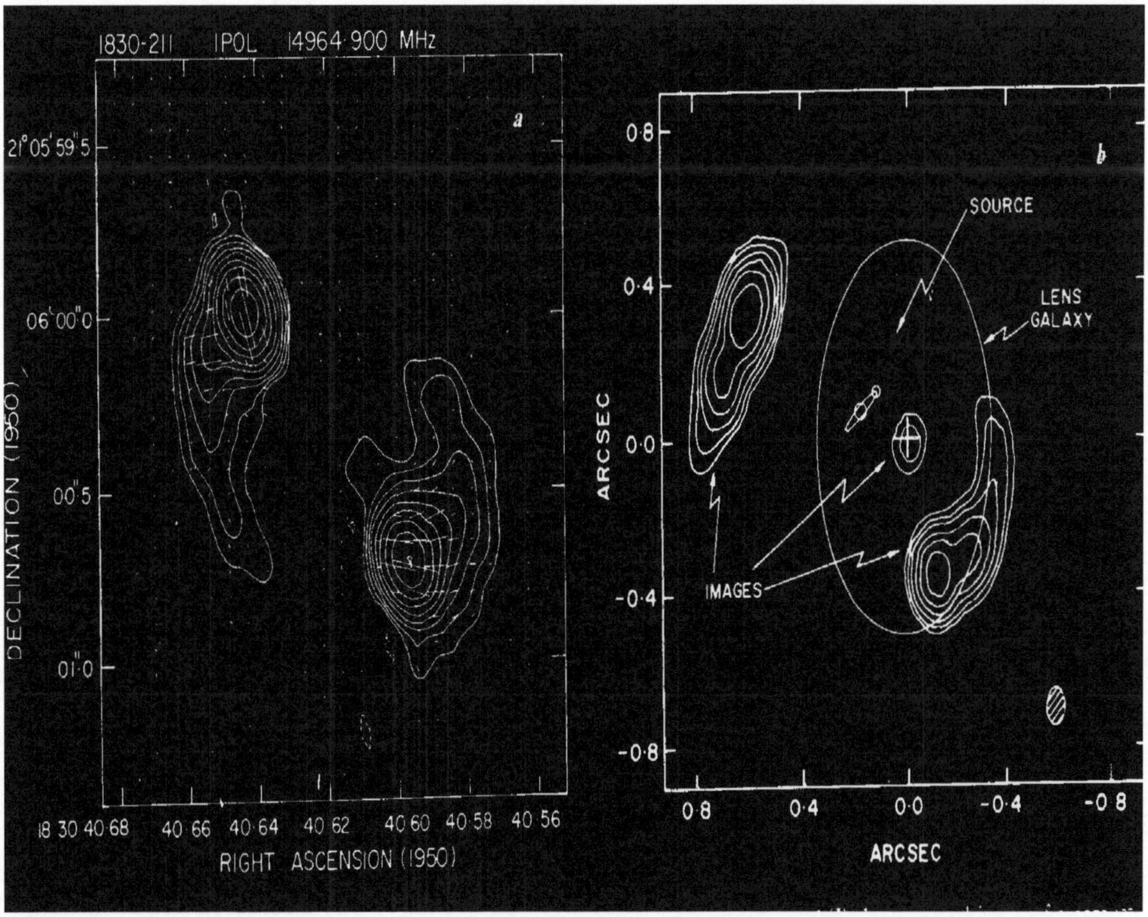

Fig. 5.10 The left figure shows the observed map of the double flat spectrum radio source made with the VLA at 11 GHz. Right Figure shows a derived image of a distant radio quasar with a core and a jet lensed by an intervening galaxy (Subrahmanyan et al. 1990)

The Giant Radio galaxy 0503-286: This radio source with a size of 2.5 Mpc (Figure 5.11) is one of the largest known radio galaxy in the Southern Sphere and was independently discovered using the OSRT by Saripalli et al. (1986) and with the Molongolo Synthesis Telescope (Subrahmanya and Hunstead 1986) Observations of the Giant Radio Galaxies provide information about the luminosity of the nuclear regions of these galaxies, inverse Compton scattering and constraints on the physical properties of the intergalactic medium.

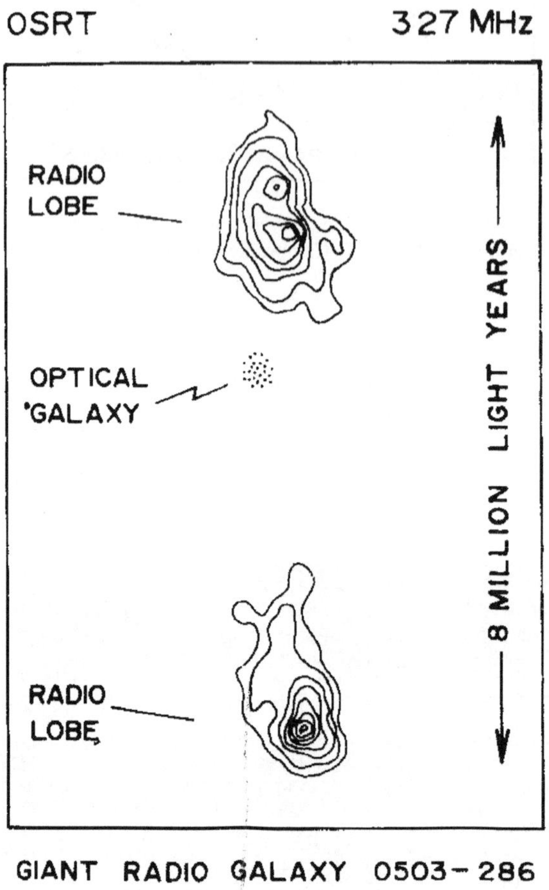

Fig. 5.11 The Giant Radio galaxy 0503-286 discovered in 1986 is nearly 8 million light years across (Saripalli et al. 1986)

Observations of Low Frequency Variables: A well-defined sample of 100 flat-spectrum radio sources was monitored over several months at 325 MHz. It was found that the low frequency variability supported RISS model; a relation giving dependence of the variability with the galactic latitude was derived (Figure 5.12; Ghosh 1990).

An Ultra Steep Relic Source in Abell 85: As part of an extensive study of clusters of galaxies, an ultra steep spectrum radio source without any obvious optical counterpart was discovered in Abell 85 (Figure 5.13) (Joshi et al. 1986). This source has evinced considerable observational and theoretical interest by several workers in the last few years.

5.2.3.4 Galactic Objects

Radio emission has been observed from a variety of objects in our galaxy, ranging from the Jupiter, sun, neutral HI gas in spiral arms, non-thermal emission, recombination lines from interstellar gas, ionized hydrogen HII regions, molecular clouds, pulsating radio sources (pulsars), micro-quasars, galactic centre, etc. Radio astronomers in India have made observations of most of the above objects using metre wave radio telescopes built in India and also used international facilities. Here we describe some of the results obtained with the

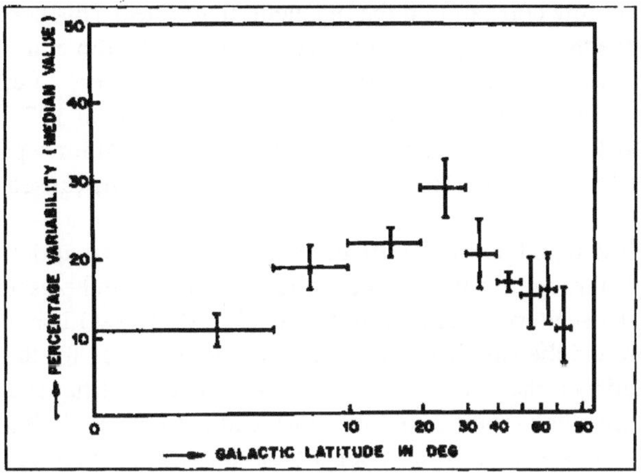

Fig. 5.12 Dependence of percentage variability of radio sources on galactic latitude by RISS

ORT by the TIFR group. In Section 5.3 are given observations made with the GMRT and in Sections 5.4, 5.5 and 5.6 are some of the results obtained by IIA, RRI and PRL.

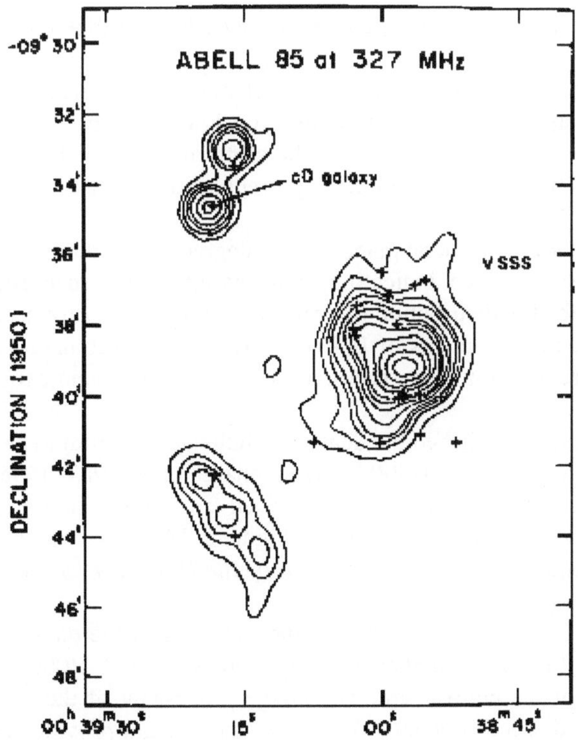

Fig. 5.13 The relic source with very steep spectrum (VSSS) discovered in 1986 with the OSRT (Joshi et al. 1986) has been studied in detail by many researchers in recent years

The Galactic Centre: The radio source Sgr-A associated with the centre of our galaxy is one of the strongest radio source in the sky. The Ooty occultation observations of Sgr-A at 325 MHz were combined with the WSRT observations at 1,420 MHz yielding separation of the thermal and non-thermal emission within Sgr-A and clarifying the properties of this important radio source for the first time. It was shown that the non-thermal emission traced a shell-like structure with a diameter of $\sim$ 8 pc, marked by several peaks superimposed on a broader non-thermal source of flatter spectra (Gopal-Krishna et al. 1972; Gopal-Krishna and Swarup 1976).

Pulsar Research with the ORT: Pulsars are rapidly rotating neutron stars in which the radio emission arises by curvature radiation in a narrow cone around the magnetic field lines. This cone sweeps past the observer like a lighthouse beacon. Thus the observer receives a narrow pulse of radio emission during each rotation of the neutron stars. Observations with the ORT during 1970s led to discovery of eight new pulsars, including a pulsar with unusually large pulse-width (Mohanty and Balasubramanian 1975). Detailed observations of several pulsars were made by Krishnamohan and Balasubramanian (1984) for separating the degree of intrinsic intensity variations of the pulsed emission from that due to fluctuations caused by the interstellar medium (ISM).

Using the upgraded ORT, detailed pulsar scintillation studies were carried out during 1993–1998. Propagation of pulsar radio signals through the turbulent plasma of the ISM can be used for a better understanding of the ISM. It was shown that scattering material in the local ISM (LISM) is not uniformly distributed in a region of about 1 km/sec around the sun. Instead, there is strong evidence for the presence of a low-density bubble around us, surrounded by a shell of enhanced scattering material (Bhat et al. 1998). This result was an important step in understanding the properties of the LISM, and has been included in electron density models of the galactic ISM. It was also shown that the Loop I bubble (which is believed to be a nearby supernova remnant shell) also produces enhanced scattering from the plasma associated with its shell boundary (Bhat and Gupta 2001).

Backer (1970) discovered that some pulsars occasionally show drop in their pulsed emission, called nulling. It provided very tight constraints for their emission physics. Using the ORT, pulse nulling was studied for 10 pulsars. Nulling fractions for PSRs B0149-16 and B0942-13 were determined in detail (Vivekanand 1995). A high sensitivity study of PSR B0031-07 was carried out to characterize its unique sub-pulse drifting. The ORT data showed that the average spacing between two sub-pulses increases monotonically with drift rate contrary to the belief held earlier (Vivekanand and Joshi 1997; Joshi and Vivekanand 2000).

5.2.3.5 Interplanetary Scintillations (IPS) and Solar Weather Studies Using the ORT (1970–2008)

During the last 38 years, the Ooty group has made important contributions concerning fine structures of radio sources, solar wind and interplanetary medium based on interplanetary scintillation (IPS) observations of distant radio sources. The IPS technique exploits the scattering of radiation from compact components of quasars and radio galaxies by the solar wind density irregularities, which are moving radially outward from the sun.

An extensive IPS survey of the galactic plane ($b < 10$ degree) using the ORT showed absence of scintillating radio sources of size <0.5 arc-sec, indicating enhanced interstellar scattering towards the galactic plane. A two-component model for the distribution of scattering plasma was proposed (Rao and Ananthakrishnan 1984). Further, systematic measurements of IPS on individual sources have yielded information about the compact component of size $\leq$100 mill-arc-sec (mas) for several thousand extragalactic radio sources at $\sim$1 m wavelength.

Ooty IPS studies have also established the density turbulence spectrum to be of power-law form, $\Omega_{Ne}(q) \sim q^{-(3to3.5)}$ in the spatial-scale range 10–500 km (Manoharan et al. 1987, 1995 and 2000). Using the above power-law model, the speed of the solar wind was derived from single-station measurements that were consistent with the multi-station IPS system in Japan (Figure 5.14) (Manoharan and Ananthakrishnan 1990). *This is a very important result and has been accepted by the IPS observers who are using the single-station method for deriving solar wind velocity.*

The importance of the Ooty IPS measurements increased when the day-to-day monitoring of the heliosphere was made on a grid of large number of radio sources ($\sim$ 800–1,000 per day), whose lines of sight cut across different parts of the heliosphere. The image processing of the normalized scintillation indices (g-values) and the estimated speeds obtained from each line of sight provides the three-dimensional view of the ambient solar wind (Figure 5.15) as well as the turbulent regions associated with the propagating disturbances,such as coronal mass ejections (CMEs) in the IPS field of view (e.g. Manoharan et al. 1995; Janardhan et al. 1996).

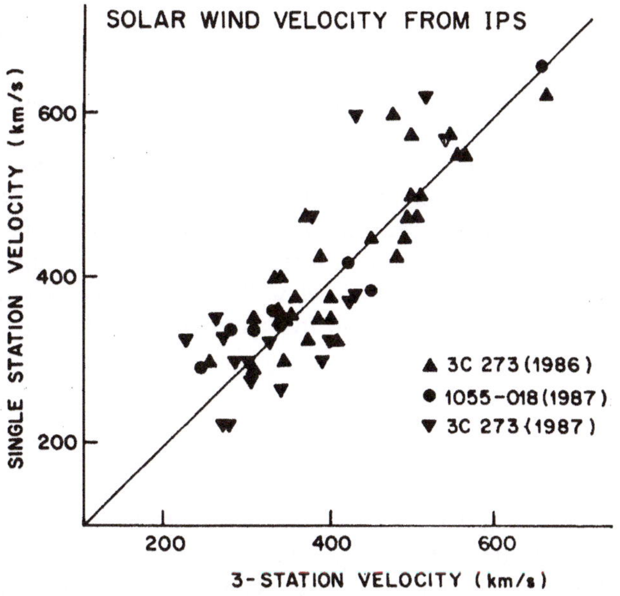

Fig. 5.14 Plot of solar wind velocity derived from single station IPS power spectra observations made with the ORT versus 3-station IPS system in Japan (Manoharan and Ananthakrishnan (1990)

The Ooty IPS studies have provided for the first time a complete coverage of the imaging of CME structures all the way from the sun to the earth (Manoharan et al. 2001). The velocity of the solar wind is found to have two-level deceleration: (1) a low decline in speed at distances within or about 100 R_{sun}, and (2) a rapid decrease at larger distances from the sun. Each CME tends to attain the speed of the ambient solar wind at 1 AU or further out of the earth's orbit (Manoharan 2006). The solar wind measurements at Ooty over the solar cycle 23 provided the large-scale changes of latitudinal features of the solar wind density turbulence and speed. Routinely imaging of solar wind disturbances at Ooty in the sun-earth distance has provided a capability to predict the adverse space-weather events before their arrival at the near-earth environment (Manoharan et al. 2001; Manoharan 2006).

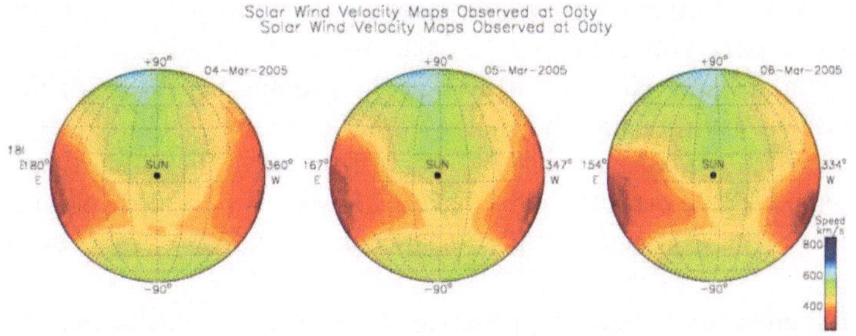

Fig. 5.15 Speed of the solar wind (V-map) on three consecutive days (Manoharan 2006)

5.3 Astronomy with the GMRT during 1999–2008 (Some Highlights)

GMRT of the National Centre for Radio Astrophysics of the Tata Institute of Fundamental Research (NCRA-TIFR) started partial observations in 1999 and was fully commissioned in 2000 for use by astronomers in India. Since 2001, its usage is decided by a GMRT time allocation Committee against refereed scientific proposals submitted by the Indian and international astronomical community. GMRT is being used by astronomers from 22 countries. Based on refereed proposals, about half of time of the GMRT has been allocated each to Indian and international astronomers. Some of the highlights of the contributions made by the Indian astronomers during the last 10 years are described below. The broad categories are cosmology with the 21 cm line of neutral atomic hydrogen, pulsars, and low frequency probing of active galaxies.

5.3.1 Cosmology

Damped Lyman Alpha Systems: Even the partially complete GMRT had the sensitivity and frequency coverage to detect absorption lines produced by intergalactic clouds against background continuum sources. In fact, its unique frequency coverage allowed one to detect sources that were not observable at other observatories. The first of these studies started in 1999 when only eight of the 30 GMRT antennas were available (Chengalur and Kanekar 1999), and the GMRT continues to be one of the foremost instruments for such studies. Observations of these systems (the so-called "damped Lyman alpha" systems, which are believed to be the precursors of the large disc galaxies we see today) have established a broad trend of increase in gas temperature with increasing look back time (e.g. Chengalur and Kanekar 2000; Kanekar and Chengalur 2003). The GMRT was also one of the first telescopes at which surveys for absorption from the OH molecule at cosmological distances have been conducted. These studies (e.g. Kanekar and Chengalur 2002), have shown that the gas phase chemistry in cosmological distant galaxies is similar to that in our own. Further accurate comparison of the frequency of the redshifted OH spectral lines, with the frequencies measured in terrestrial laboratories has helped to place stringent limits on the evolution of the value of the fine structure constant and the electron to proton mass ratio (Chengalur and Kanekar 2003).
HI Absorption from Cold Gas at Intermediate Redshifts: The Damped Lyman alpha systems (DLAs), with log N (HI) >20.3, are a major reservoir of HI at high z and possibly the progenitors of present-day galaxies. At high z, despite many attempts, only a handful of DLA galaxies have been detected based on line and/or continuum emission. A GMRT survey to search for 21 cm absorption in a representative and unbiased sample of 35 DLA candidates at $1.10 < z < 1.45$, drawn from the strong MgII systems in SDSS DR5, has resulted in discovery of nine new 21 cm absorbers. Prior to this survey, only one 21 cm absorber was known in the redshift range of $0.7 < z < 2$ (Gupta et al. 2006b). GMRT observations of two relatively weak, radio loud, but very dusty QSOs resulted in the detection of 21 cm absorption in both cases. The spin temperature of the gas is of the order of or smaller than 500 K (Srianand et al. 2008).

Search for HI Emission: Observations of neutral hydrogen in individual galaxies at different cosmic epochs provide a unique probe of star formation and the assembly of galactic discs. However, this is severely limited even for telescopes with a large collecting area like the GMRT. In a statistical study, however, one can add the signals for many galaxies and gain information on the average behaviour of the population, and begin to examine dependence on environment and galaxy type. The GMRT is uniquely suited to such studies, and indeed the highest redshift existing constrains come from GMRT observations (Lah et al. 2007). The observations of field galaxies give measurements that are consistent with those obtained via other techniques. However, measurements of the gas content of galaxies in clusters indicate a rapid evolution of the gas content of galaxies in clusters.

Dwarf Galaxies: It is currently believed that galaxies formed hierarchically, that is, small objects collapsed first, and these in turn merged to form the large galaxies that we see around us today. This process is inherently inefficient, leaving every large galaxy surrounded by a host of unmerged smaller (or "dwarf ") galaxies. The nearby dwarf galaxies can hence be regarded as representative of the primordial galaxy population and can be studied in exquisite detail. As some of the most un-evolved systems in the nearby universe, observations of dwarf galaxies are relevant in a host of cosmological contexts, ranging from testing predictions of cold dark matter models, the influence of cosmic reionization on the baryon content of galaxies, to understanding the parent populations of quasar absorption line systems. Further, dwarf galaxies also provide unique sites for understanding the processes which govern the conversion of gas to stars. This is because, (i) as opposed to large spiral galaxies, dwarf galaxies are dynamically simple systems and (ii) the relatively pristine chemical composition of gas in dwarf galaxies is closer to what one would expect in primordial galaxies. The largest such study is based on a GMRT survey of FIGGS, (the Faint Irregular galaxy GMRT Survey by Begum et al. 2008, and references therein). The GMRT studies have revealed ordered kinematics with dark matter profiles that do not match the expectations of cold dark matter that are substantially different from that observed in large galaxies simulations (e.g. Begum et al. 2003), and star formation rates. The GMRT map of the HI of the dwarf irregular galaxy NGC 3741 (Figure 5.16) extends to a record 8.3 times the Holmberg radius (Begum et al. 2005).

Fig. 5.16 The GMRT map of the HI, overlaid on the optical image of the dwarf irregular galaxy NGC 3741 ($M_B \sim -13.1$). HI disk extends to a record 8.3 times the Holmberg radius. The rotation curve is measured up to 38 optical scale and remains flat; with a dynamical mass-to-light ratio of 107, this is one of the darkest dwarf irregulars (Begum et al. 2005)

5.3.2 Active Galaxies

Active galaxies could be broadly classified into those which harbour an AGN or exhibit an intense burst of star formation, called the starburst galaxies. The AGNs were amongst the first to show strong radio emission and are now understood to be due to accreting supermassive black holes pouring out energy in the form of jets, whose origin and formation remain enigmatic, into their host galaxies and beyond, into the intergalactic medium. Low-frequency data with the good sensitivity and resolution provided by the GMRT has led to both interesting individual objects and statistical information about the population. In both cases, the multi-frequency capability has provided valuable information.

Double-Double Radio Galaxies: One of the important issues concerning galaxies is the duration of their AGN phase and whether such periods of activity are episodic. For the radio-luminous objects, an interesting way of probing their history is via the structural and spectral information of the lobes of extended radio emission. A striking example of episodic jet activity is when a new pair of radio lobes is seen closer to the nucleus before the 'old' and more distant radio lobes have faded. Such sources have been christened as 'double-double' radio galaxies (DDRG). Observations made with the GMRT have led to the discovery of a new and interesting DDRG (J1453+3304, Figure 5.17), (Saikia et al. 2006), while observations of a number of DDRGs over a large frequency range have led to estimates of spectral ages and hence time scales of episodic activity (e.g. Konar et al. 2006).

Giant Radio Galaxies: The discovery of the largest giant radio galaxy, which are defined to be over a Mpc in size, has been reported from observations made with the GMRT and other telescopes (Machalski et al. 2008), and also one of the largest one-sided radio jets (Figure 5.18) (Bagchi et al. 2007). Multi-frequency matched-resolution observations of such giant radio sources (e.g. Konar et al. 2008), X-shaped radio sources

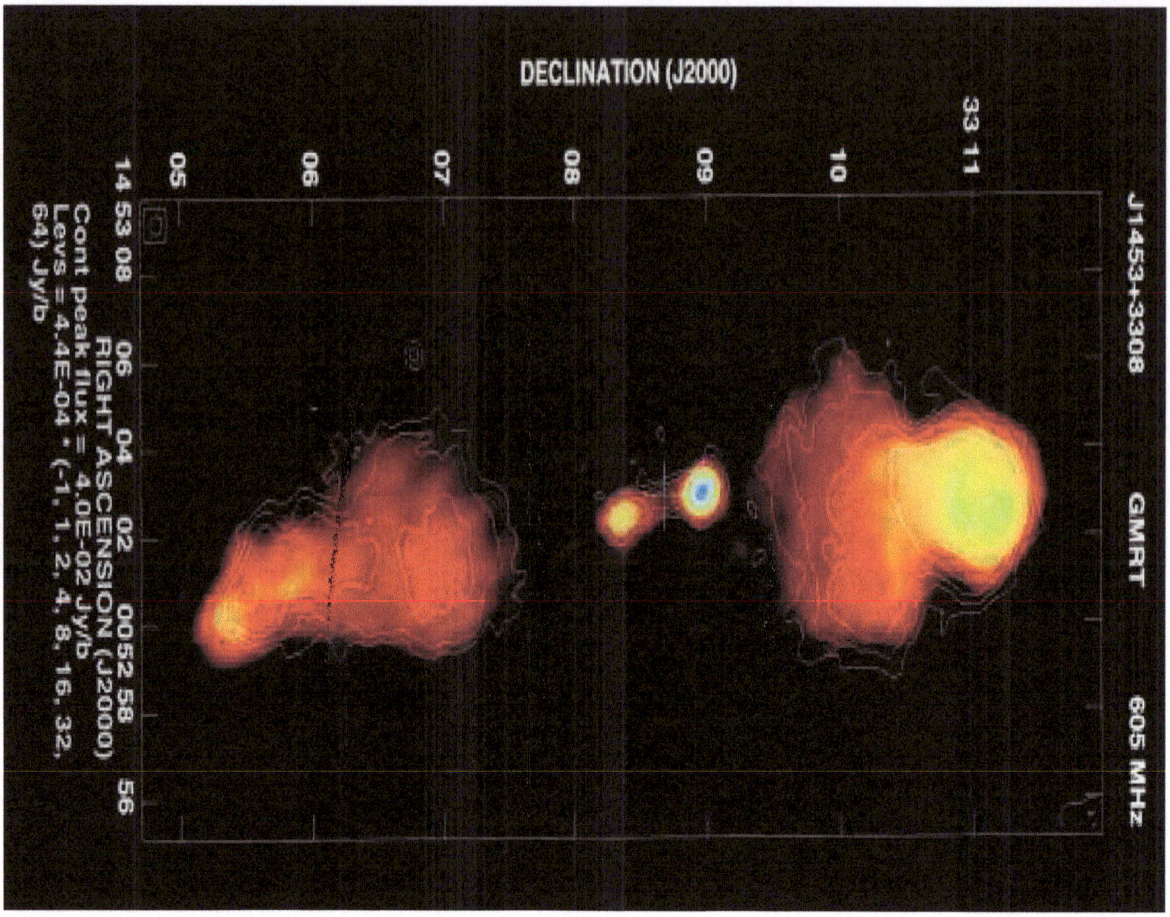

Fig. 5.17 Brightness distribution across the double-double radio galaxy J1453+3304

(Lal and Rao 2007), relic lobes of radio sources have led to spectral and age information about different parts of the source, constraining models of radio sources.

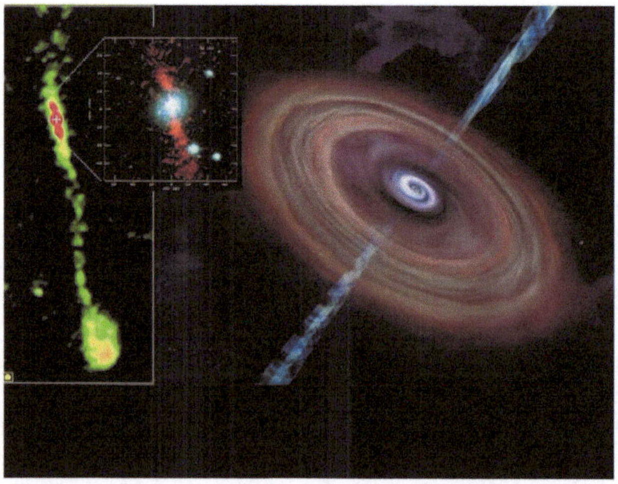

Fig. 5.18 Radio image of a very large one-sided jet discovered by Bagchi et al. (2007)

Radio Relics and Halos in Clusters of Galaxies: Radio halos and relics are low-surface brightness sources with steep radio spectra, and sizes which can extend to over a Mpc.

While radio halos are usually located near the centre of a cluster and have a reasonably regular structure, the relics are located in the periphery of the clusters and exhibit a wider variety of structures. Besides understanding the physics of these structures these could provide useful insights towards understanding formation and evolution of clusters of galaxies. GMRT observations have helped identify several new relics and halos, and low-frequency GMRT observations of these have helped constrain models of these objects (e.g. Brunetti et al. 2008; Venturi et al. 2009).

Interacting Galaxies and Superwinds: In active galaxies, the gaseous ISM of disk galaxies are affected by various energetic and violent phenomena such as supernovae, stellar winds, jets driven by AGN, winds from accretion-disk and ram pressure stripping. Superwinds from starburst galaxies are important in supplying metal-enriched gas to the halo of the disk galaxy as well as to the intergalactic medium or the intracluster medium. Several such systems have been studied in detail using the GMRT providing important and useful constraints on the models of these sources, such as the superwind galaxy NGC1482 (Hota and Saikia 2005); in the Seyfert galaxy NGC3079 a large-scale halo has been imaged with the GMRT (Irwin and Saikia 2003); also the highly disrupted Seyfert galaxy in the Virgo cluster NGC4438 (Hota et al. 2007). The member galaxies in Holmberg 124 display signatures of tidal interwction and ram pressure stripping in the GMRT radio continuum and HI images (Figure 5.19) (Kantharia et al. 2005). These lower-luminosity galaxies also harbour black holes, somewhat more massive than in our own galaxy.

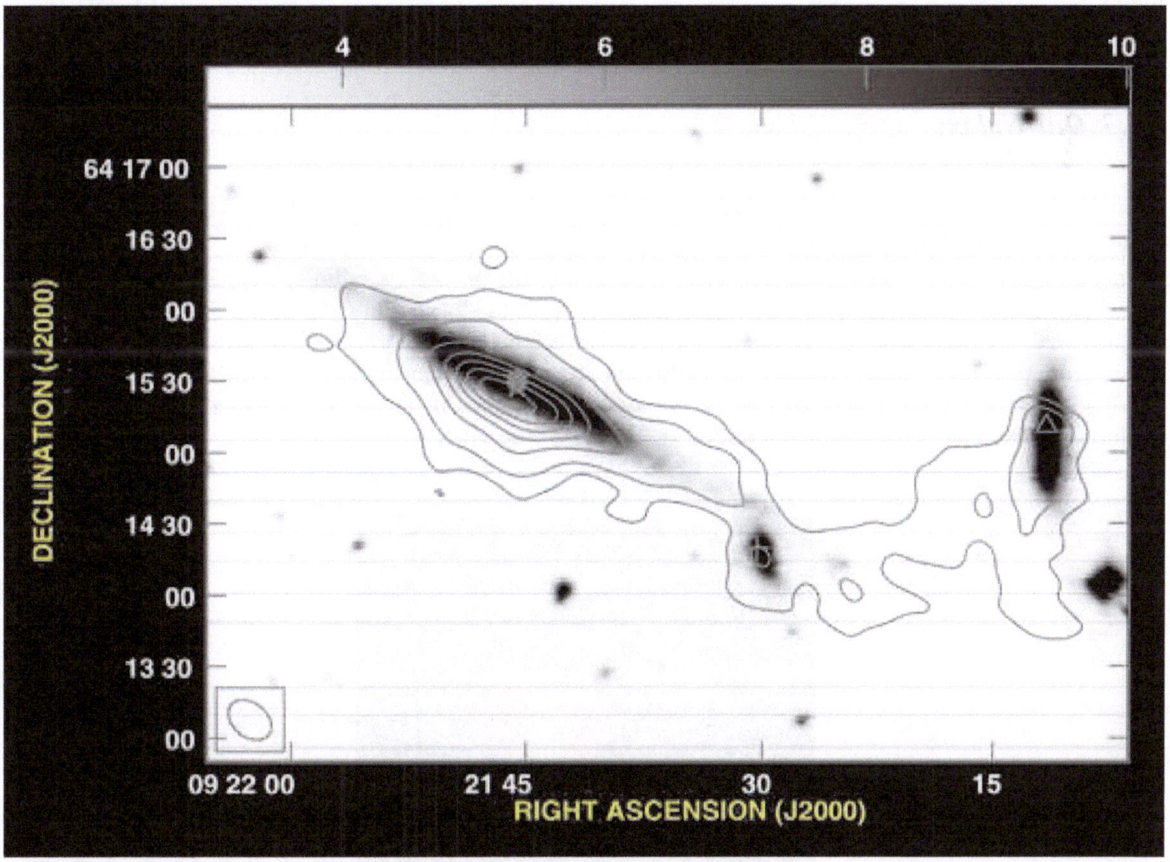

Fig. 5.19 GMRT image of three galaxies in the group Holmberg 124, NGC2820, NGC2814 and Mrk 108, superimposed on optical image in grey scale. The radio bridge with a very steep spectrum (α = -1.8) is likely to be due to interactions of the galaxies, whereas the displaced radio disk of NGC2814 is likely due to ram pressure stripping (Kantharia et al. 2005)

Gamma Ray Bursts: GMRT has played key role in identifying radio counterparts of very high energy gamma ray sources at high redshifts. Deep GMRT radio observations of these energetic sources have for the first time revealed presence of extended, diffuse and compact radio sources which are positionally coincident with the X-ray and optical/IR sources (Paredes et al. 2007).

HI Absorption Studies: An understanding of the properties of the gaseous environments of radio galaxies and quasars could provide valuable insights towards understanding the phenomenon of radio activity associated with these objects and their evolution. Such studies also enable us to test consistency of these properties with the unified schemes for these objects. An important way of probing the neutral component of this gas over a wide range of length scales is via 21 cm HI absorption towards radio sources of different sizes. As part of a study of HI absorption towards the central regions of active galaxies using the GMRT, HI absorption lines have been discovered towards compact sources, the core of a radio galaxy and a rejuvenated radio source, consistent with the suggestion that supply of fresh gas restarts the radio activity (Gupta et al. 2006a; Gupta and Saikia 2006; Saikia et al. 2007).

Deep Low Frequency Surveys: Radio sources stronger than a few mJy are usually associated with AGN, while at lower flux densities the radio source counts are dominated by the radio-quiet objects in early-type galaxies, low-luminosity AGN and contributions from starbursts in late-type galaxies. In the last few years, the capability of doing deep low frequency imaging has been greatly improved by careful study of systematics and elaborate data processing. The deep surveys of specific fields at 610 MHz (Garn et al. 2008), 325 MHz (Sirothia et al. 2009a, 2009b) and 150 MHz (George and Ishwara-Chandra 2009) have reached depth and resolution unprecedented in these bands. Comparisons with existing high frequency surveys now become meaningful and they are already revealing interesting classes of objects with steep, curved, or peaked radio spectra, and unusual morphologies, for further follow-up. In many cases, extensive information at other wavebands is also available, allowing for a multi-wavelength approach to modeling the population of active galaxies. The surveys carried out so far represent the tip of what is possible and quite likely in the immediate future.

5.3.3 Our Galaxy

Many investigations have been done concerning objects in our galaxy over the last 8 years by research workers in India and abroad. Only a few of the results are described here.

Galactic Centre region: The lowest frequency detection of the black hole radio source, Sgr A* at the centre of our galaxy at $\sim$ 620 MHz with the GMRT indicated that it is located in front of the Sgr-A west complex (Figure 5.20) (Roy and Rao 2004). Scatter broadening has been found to occur towards the galactic centre region from the GMRT observations of 26 compact extragalactic radio sources at 255 MHz and 154 MHz. It was also inferred that the 7’ halo in Sgr A complex is a non-thermal source rather than a mixture of thermal and non-thermal electrons (Roy and Rao 2006).

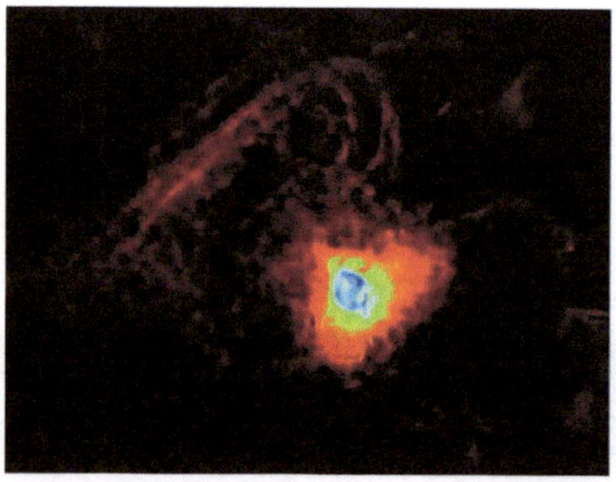

Fig. 5.20 A radio image of the galactic centre and surroundings made with the GMRT at 610 MHz (Roy and Rao 2007)

Micro-quasars and Gamma-ray Sources: Micro-quasars in our galaxy have been monitored regularly using the GMRT since 2002. Mini flares were observed from one of the famous micro-quasar GRS1915+105, using which a new method to obtain radio spectral index from single frequency radio observation was evolved through modeling of evolution of flare under adiabatic expansion (Ishwara-Chandara et al. 2002). Further observation of GRS1915+105 during the flare also showed that the radio emission is optically thick at low radio frequencies in the initial stages of the flare. This was the first extensive monitoring of a micro-quasar at metre wavelengths. The flare peaks at metre wavelengths with a delay of a few days from its peak at cm

wavelengths (Pandey 2006). Forty newly discovered gamma-ray sources by INTEGRAL were observed with the GMRT that led to the discovery of four new micro-quasar candidates (Pandey 2006).
Novae: GMRT has been used to study the radio emission from galactic cataclysmic binaries known as novae which consist of a white dwarf accreting matter from a star with a large envelope. (i) The classical nova GK Persei was imaged at 325 and 610 MHz (Anupama and Kantharia 2005) Its flux density was found to follow a spectrum with index -0.85 whereas the higher frequencies had a flatter spectrum (-0.7) indicating an adiabatic expansion of the nova remnant. (ii) RS Ophiuchi, a recurrent nova which recorded its latest outburst in February 2006 was observed at frequencies of 240, 325 and 610 MHz (Kantharia et al. 2007). This was the first detection of a recurrent nova at frequencies below 1 GHz and clearly demonstrated the presence of non-thermal synchrotron emission of index -0.8 in this system at very early times.
Sun: High dynamic range snapshot images of the solar corona were made by combining visibilities from the GMRT and Nancay radio heliograph in France. The technique allowed obtaining images of the sun at 327 MHz with a resolution of 49 arc-sec up to the size of the whole sun (Mercier et al. 2006).

5.3.4 Pulsars

Due to their steep spectrum, pulsars have higher flux densities at lower frequencies. Further, the large collecting area of the GMRT allows observations of individual pulses with good sensitivity. Observations of individual pulses in nearly aligned pulsars have revealed details of drift and polarization patterns that provide important clues to the still enigmatic emission mechanisms. A few notable pulsars have also been discovered using the GMRT.
Emission Geometry of Radio Pulsars: From detailed analysis of high quality single pulse data taken primarily with the GMRT, new results have been obtained about the detailed distribution of emission components in the profiles of several pulsars (Gangadhara and Gupta 2001; Gupta and Gangadhara 2003). These results support the model of concentric conal rings of emission around a central core component of emission (Figure 5.21). Further, slight asymmetry in the locations of the conal components can be interpreted as being due to retardation and aberration effects in the pulsar magnetosphere. The main conclusions are (i) pulsars with multi-component profiles have multiple (two to three) concentric, hollow cones of emission, (ii) the typical emission heights for these cones in the pulsar magnetosphere range from $\sim$ 100 km to $\sim$ 1000 km, (iii) emission height for or a given cone is a function of the radio frequency, being lower for the higher frequencies, (iv) the magnetic field lines associated with these emitting cones are not located at the edge of the open field line region, but lie in the range $\sim$ 0.2–0.7 of this boundary, (v) there is some evidence that the wider cones originate further out on field lines.
Multi-frequency Pulsar Observations: Detailed studies of pulsars have been carried out using multi-frequency observations, simultaneous or otherwise, done at the GMRT in combination with other radio telescopes. These have yielded interesting results in diverse topics such as (i) estimation of pulsar dispersion measures (Ahuja et al. 2005), (ii) emission geometry of the drifting pulsar PSR B0031-07 (Smits et al. 2007), (iii) wide-band nulling properties of radio pulsars (Bhat et al. 2007), (iv) frequency dependence of pulse broadening due to interstellar scintillations (Loehmer et al. 2004). Furthermore, when combined with polarization information, extra insight into the emission process has been achieved, as in the case of the emission from the central core region of PSR B0329+54 (Mitra et al. 2007).
Study of Drifting Pulsars: Sensitive data from the GMRT on a few pulsars with wide emission profiles has provided interesting new insights (Gupta et al. 2004; Bhattacharyya et al. 2007). PSR B0826-34, a pulsar with emission over almost the whole pulse period, has been shown to have as many as seven drift bands in the main pulse window with highly correlated variations over durations of tens of pulse periods. This is first direct evidence for such a large number of sparks and provides strong support to the model of rings of circulating sparks on the polar cap, resulting in conal emission beams. The unique drift pattern in the wide profile pulsar, B0818-41, is modeled as being created by two conal rings. Both the rings have the same values for the pattern rotation period (18.3 pulsar periods) and the number of emission sparks (about 20). GMRT has also allowed unraveling of the sub-pulse modulation characteristics of the five component pulsar B1857-26 (Mitra and Rankin 2008). The outer conal components exhibit a modulation periodicity of 7.4 pulsar periods, and there is evidence to indicate that there are 20 sparks and the pattern rotation period is 147 periods.
Studies of the Double Pulsar: The double pulsar system PSR J0737-3039, discovered by Lyne et al. (2004), provides an excellent laboratory for relativistic physics and studies of pulsar emission mechanism. The system is unique due to sharp pulses of this msec pulsar, the detection of its long period companion as a pulsar and

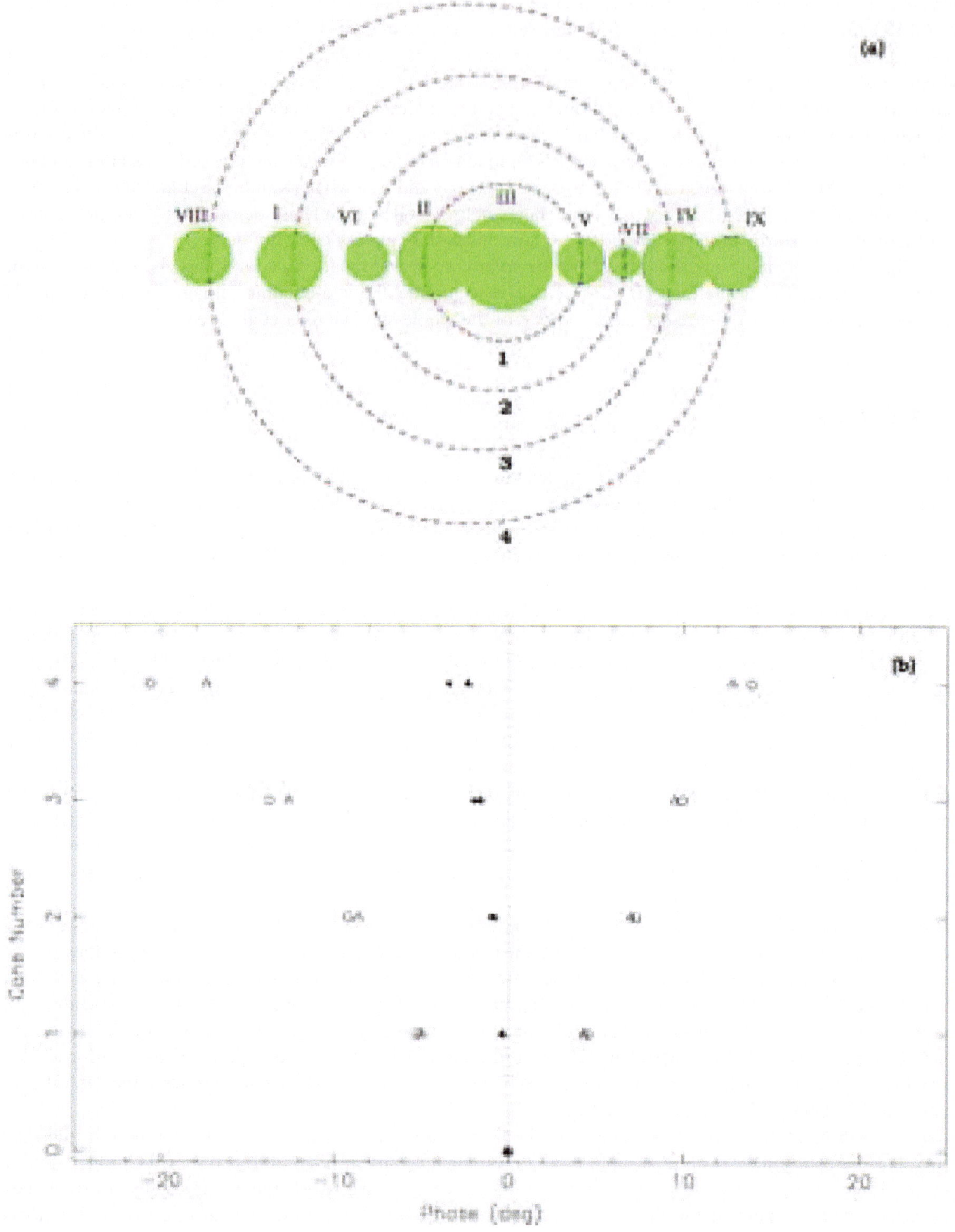

Fig. 5.21 Location of nine emission components detected for PSR B0329+54 at 610 MHz, shown in the form of four conal rings around a central core component (top panel). The core component is labeled as cone. Lower panel illustrates the increasing asymmetrical locations of the four conal component pairs, with respect to the core, for 610 MHz (triangles) and 325 MHz (circles), (Gupta and Gangadhara 2003)

an almost edge-on orbit. The line of sight to the msec pulsar passes through the magnetosphere of the long period pulsar giving a unique probe of the pulsar magnetosphere. Multi-epoch observations with GMRT at 325 MHz of the long period pulsar in this double pulsar system showed a change in the intensity, duration, and separation of its two bright phases, which provides a unique probe for the geodetic precession of this pulsar, with implications for theories of relativity (Joshi et al. 2004; Joshi 2006).

Discovery of Some Interesting New Pulsars Using the GMRT: Although more than thousand pulsars are now known, some of those discovered with the GMRT are of special interest. (i) The first pulsar PSR JO514-4002A discovered by the GMRT was found to be a 4.99 msec period neutron star, in the globular cluster NGC 1851 (Freire et al. 2004). Detailed follow-up studies revealed it to be a very interesting binary pulsar, with a white dwarf companion, an orbital period of 18.8 days and an orbital eccentricity of 0.9 - *the most eccentric orbit of a pulsar in a binary system.* The formation of such systems points to significant three body interactions *in the dynamics of dense globular clusters.* (ii) The second pulsar discovered by the GMRT was a very young pulsar, second *youngest pulsar in our galaxy. It is* located at the centre of the supernova remnant G21.5-0.9. The PSR J1833-1034, has a spin period of 61.8 msec, a characteristic age of about 5,000 years and a spin-down luminosity that is the second highest amongst all the known pulsars in our galaxy (Gupta et al. 2005): *given that most work on young pulsars in supernova remnants was done with statistics of two, one more is a welcome addition*; follow-up studies of this pulsar have revealed very interesting glitching behaviour (Roy et al. in preparation). (iii) An unbiased survey for long period pulsars carried out with the GMRT at 610 MHz (Joshi et al. 2008) covering 106 suare degrees has yielded its first results: three new pulsars (PSR J0026+6320, PSR J2208+5500, PSR J2218+5729 with periods of 0.318, 0.933 and 1.06, respectively). Results from this survey should improve our understanding of aging and death of these objects.

5.3.5 Astronomy with the GMRT: General Remarks

It is important to note that the above account has focused on selected highlights mostly by NCRA astronomers and is only a small subset of a much more extensive body of work, amounting to more than 300 publications done with the GMRT since it was commissioned. In particular, a large body of work on radio galaxies, quasars, cluster of galaxies, nearby galaxies, transients, ISM of our galaxy, supernova remnants, HII regions and sun is not represented. The proceedings of a recent international conference held at NCRA in December 2008 on “low frequency radio universe” (Saikia et al. 2009) describe in detail many studies carried out by Indian and international astronomers using the GMRT and other radio telescopes.

5.4 Radio Astronomy at the Indian Institute of Astrophysics

5.4.1 Solar Physics

The main focus of the radio astronomy facilities at the IIA concerns observations of sun at low frequencies, viz. 30–150 MHz. The early observations date back to 1950s at the Kodaikanal observatory, forerunner of the institute. Continuous recording of solar radio noise flux was commenced under A. K. Das in 1952 using a 100 MHz interferometer with twin Yagi type antennas (Kodaikanal Observatory report 1952). A 20-foot equatorially mounted paraboloid for observations at decimetre and metre wavelengths was set up in 1961 after M. K. Vainu Bappu took over as the director of the observatory. Under the Kodaikanal-Yale Project, recording of radio radiation from Jupiter at a frequency of 22.2 MHz was started using a phase switching interferometer (Kodaikanal Observatory report 1962). The custom-built 3 GHz wavelength radio receiver from the CSIRO, Australia, was put to use in 1965 for regular solar monitoring (Kochar and Narlikar 1995). In the early 1970s, an antenna array operating at 25 MHz was used to obtain information on radio bursts from the outer solar corona, with high temporal and frequency resolution (Sastry 1973). The above radio telescopes were discontinued after the Kodaikanal group moved to Bangalore in mid-1975 on the establishment of the autonomous IIA with the Kodaikanal Observatory as one of its constituent.

During the mid-1970s, a large radio telescope operating at 34.5 MHz was jointly set up by the IIA and the RRI, Bangalore at Gauribidanur, 80 km north of Bangalore. As described in Section 5.5.2, the telescope consists of 1,000 dipoles arranged in a ‘T’ configuration, with a 1.4 km east-west arm and a 0.5 km south arm

(Figure 5.22), (Sastry et al. 1981; Dwarkanath et al. 1982). It has been engaged in the study of radio waves emanating from sun and various galactic and extragalactic objects (see Section 4.3 for non-solar observations). The notable solar observations with the array are: two-dimensional images of radio emission from the slowly varying discrete sources in the outer solar corona; radio brightness temperature of the outer solar corona is $<10^6$ K (Sastry et al. 1983).

Fig. 5.22 An aerial view is shown of a section of the east-west arm of the Gauribidanur radio telescope operating at 34.5 MHz

Fig. 5.23 Shows a section of the south arm of the Gauribidanur radioheliograph

Not many white light coronagraphs to probe the outer solar corona were in operation during the early 1980s and also the interest in the study of coronal mass ejections (CMEs) had just begun. Further both the Culgoora radioheliograph in Australia and the Clark Lake radioheliograph in USA had closed down operations. So the IIA decided to build a radioheliograph for dedicated observations of the solar corona, simultaneously at different frequencies in the range 30–150 MHz. The Gauribidanur radioheliograph (GRH, see Figure 5.23) is in operation since 1997 (Ramesh et al. 1999). In the above frequency range the GRH provides information on the solar corona in the height range ~0.2–0.8 R_s above the solar surface, which is difficult to probe using ground based and space borne white light coronagraphs. Moreover no other radio telescope is presently operating in the above frequency. Some of the notable observations made with the GRH are:

Density and temperature of the pre-event structure of CMEs (Kathiravan and Ramesh 2005); velocity/acceleration of CMEs close to the solar surface (Ramesh et al. 2003); 'true' speed of CMEs in the three-dimensional space (Kathiravan and Ramesh 2004); estimation of parameters of the CMEs at large distances $\sim (5-40\ R_s)$ from the sun by observing angular broadening of distant cosmic radio sources (Ramesh et al. 2001); coronal electron density gradient in the $\sim 0.2-0.8\ R_s$ height range above the solar surface (Ramesh et al. 2006); radio noise storms and CMEs (Kathivaran et al. 2007); onset of CMEs and occurrence of transient dimming in the solar corona (Ramesh and Sastry 2000); plasma characteristics of radio emission associated with emerging magnetic flux from sub-surface layers of the solar photosphere (Shanmughasundaram and Subramanian 2004).

A high resolution radio spectrograph is used in conjunction with the GRH for obtaining a dynamic spectrum of transient burst emission from the solar corona. The antenna system consists of 8 log periodic dipoles. A commercial spectrum analyzer is used as the back end receiver to obtain spectral information with an instantaneous bandwidth of $\sim$ 250 kHz. The temporal resolution is $\sim$ 43 ms. The radio spectrograph and

GRH together provide spectral and positional information on eruptive activity in the solar atmosphere. The observations so far have provided clues to the location of a source region of a CME through observations of transient 'absorption' bursts (Ramesh and Ebenezer 2001) and occurrence of radio bursts associated with magneto-hydrodynamic shocks in the solar corona (Subramanian and Ebenezer 2006).

Recently an interference radio polarimeter has been built at the Gauribidanur observatory to understand the polarization characteristics of various solar phenomena. Based on theoretical formulations for the response of a correlation telescope to polarized radiation, an east-west one-dimensional array of 32 log periodic dipoles has been set up to probe the coronal magnetic field in the height range $\sim 0.2 - 0.8\ R_S$ above the solar surface. The dipoles are arranged as four groups and they are oriented at 0°, 45°, 90° and 135° with respect to the terrestrial north. This helps in capturing the polarization state of the incident radiation with good accuracy. The idea is to get information on the coronal magnetic field through observations of circularly polarized radio emission from discrete sources in the corona (Figure 5.24). The spectral dependence of the observed emission in the above height range can also be obtained through multi-frequency observations. Instrumental calibration is carried out through observations of the randomly polarized sidereal radio sources (Ramesh et al. 2008).

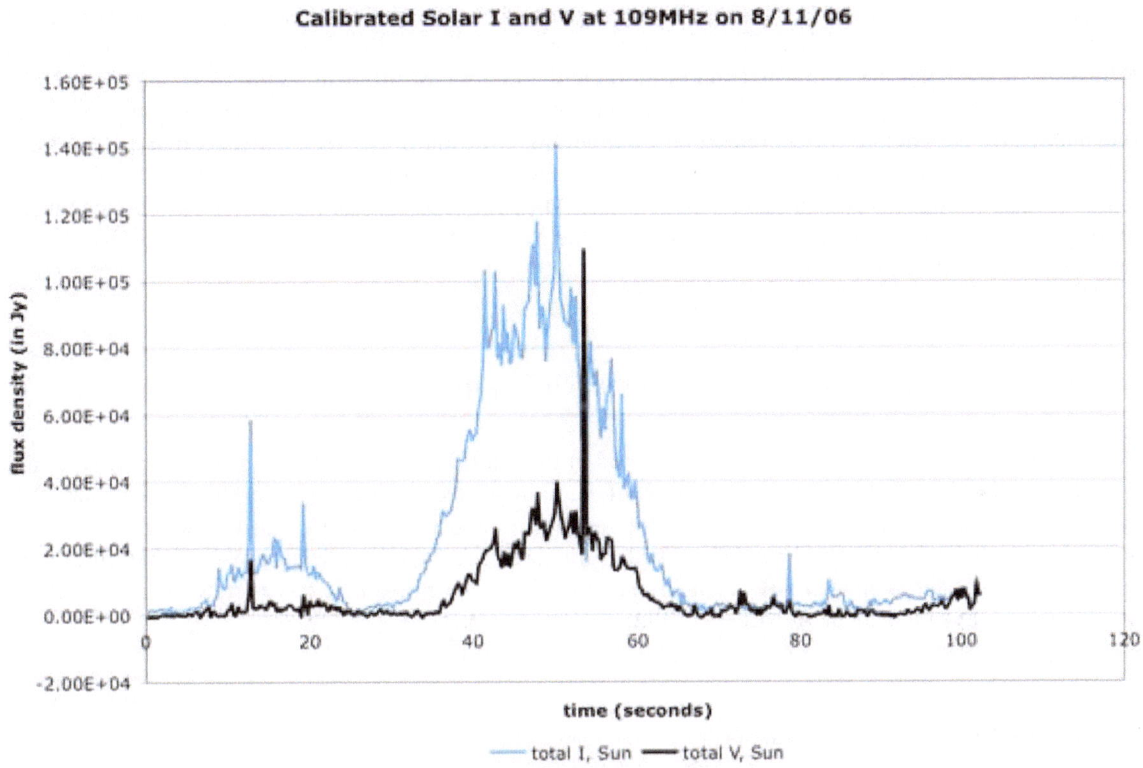

Fig. 5.24 Observations of circularly polarized radio emission from the sun at a height of about 0.4 *Rs* above the surface with the Gauribidanur radio polarimeter

5.4.2 Pulsars

Based on the data of PSR B1133+16 at three temporal resolutions (1160, 500 and 150 microseconds), it was found that orthogonal polarization modes are better resolved at higher temporal resolutions, and the pulses show a higher degree of polarizations. Further, the linear polarization is higher when the circular polarization has a minimum value (Gangadhara et al. 1999).

It was shown that the controversy whether the beam of the pulsar emission is conal or patchy has arisen because of incorrect identification of pulsar emission components in the average pulse profiles. Single pulse observations of PSR B0329+54, one of the strongest known pulsar, with the Jodrell Bank radio telescope at 606 MHz indicates that the weaker components seen in single pulses hardly show up in the average pulse profiles. Using the Window-Threshold technique for identifying the weaker emission components, nine emission components were detected in PSR B329+54 (Figure 5.21). This is the highest number of components ever detected among all the known pulsars. To make a multi-frequency estimate of the results, 325 MHz data of PSR B0329+54 from GMRT was used. The results were found to be consistent at the two frequencies. Further, it was found that the conal components on either side of the core are asymmetrically distributed with respect to the rotation phase (Gangadhara and Gupta 2001; Gupta and Gangadhara 2003).

5.4.3 *Active Galactic Nuclei*

Polarization was detected on parsec scales in the nuclei of four Fanaroff-Riley type I (low-luminosity) radio galaxies. Observations with VLBI at λ = 3.6 cm reveal the presence of ordered magnetic fields within $\sim$ 1650 Schwarzschild radii of the putative central supermassive black hole. The relatively high fractional polarization in the parsec-scale jets of these galaxies is consistent with the unified scheme for low-luminosity radio galaxies and BL Lac objects. This result also suggests that these radio galaxies lack the obscuring tori that apparently depolarize the nuclear emission in the more powerful FR II type radio galaxies, and that their supermassive black holes are poorly fed and/or inefficient radiators (Kharb et al. 2005).

Investigation was made of the point-like optical nuclei in the centres of the host galaxies of a majority of radio galaxies by the Hubble Space Telescope. Simple model-fitting of the data suggests that the emission may be coming from a non-thermal relativistic jet. The results are broadly in agreement with the unified scheme for radio-loud AGNs (Kharb and Shastri 2004). Milli-arc second scale resolution was obtained using very long baseline interferometry (VLBI) images of the Seyfert 1 and Seyfert 2 galaxies at 5 GHz to test rigorously predictions of the unified scheme (Lal et al. 2004).

5.5 Radio Astronomy at the Raman Research Institute

Radio astronomy at the RRI, Bangalore was initiated in the late 1970s, and has now grown to form a major part of the institute's activity. By this time the ORT had already been established by the TIFR group. RRI involved itself in the use of the ORT primarily for spectral line observations. One of the first results of this was an upper limit to the average interstellar Deuterium abundance. Using a frequency-switched filter bank receiver at ORT, Anantharamaiah and Radhakrishnan (1979) determined that the interstellar Deuterium to hydrogen ratio is less than about 58 parts per million.

5.5.1 *Radio Recombination Lines*

The spectral line activity gradually grew into a major enterprise at RRI in the early 1980s, and the focus shifted to Radio Recombination Lines (RRL). These lines originate in transitions between highly excited states of hydrogen and metal atoms. Typically such states are populated in the presence of an ionizing photon field. The atom is first ionized, and the high Rydberg states are populated as electrons recombine with the ions, and hence the name "recombination lines". Radio recombination lines can be excellent probes of density and temperature of these ionized regions.

An extensive survey of the galactic plane for recombination line emission around 327 MHz was undertaken by RRI using the ORT. The typical transition observed was between Rydberg levels of 273–272 of the hydrogen atom. Observations were carried out for over 3 years, using a 128-channel one-bit autocorrelation spectrometer built at RRI and installed at Ooty (Antharamaiah 1985a). The result of this survey showed that the distribution of the RRL emission is quite ubiquitous over the inner galaxy (galactic ridge), and not just associated with the known, discrete ionized hydrogen regions. As the high Rydberg level atoms can exist in only relatively low-density regions, one way to explain the widespread distribution of the RRL was that ionized hydrogen regions had hitherto unseen low-density envelopes that give rise to these lines (Anantharamaiah 1985b, 1986). Correlation of the line intensities with those of background continuum indicated significant contribution of stimulated emission caused by population inversion at these energy levels. A survey of recombination lines in the longitude range $l = 330°$ to $89°$ by Roshi and Anantharamaiah (2001) constrained the density of gas in the range 1–10 cm^{-3}.

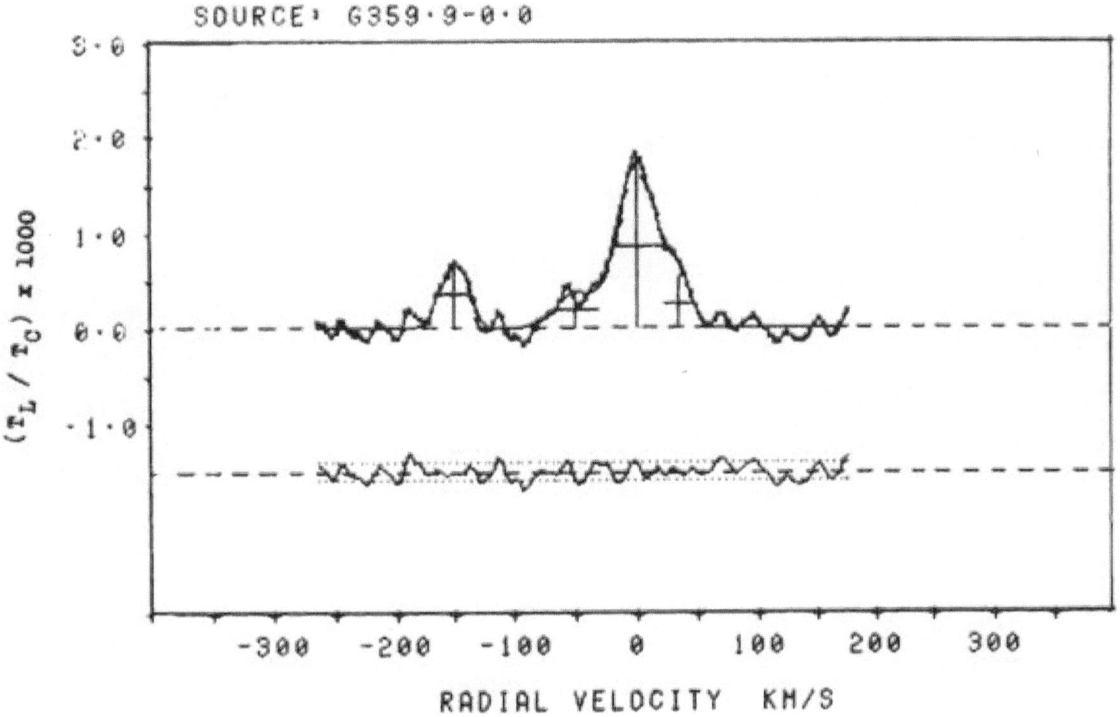

Fig. 5.25 Radio recombination lines towards the galactic centre source Sgr A detected using the ORT. The height and the width of the fitted Gaussian components are indicated using the vertical and horizontal lines, and the post-fit residuals are also shown. The feature located near −150 km/s is a recombination line of carbon while the other features belong to hydrogen (Anantharamaiah 1985a)

The strong RRL detected towards the galactic centre allowed detailed analysis and modeling of the emitting gas, and resulted in placing a lower limit to the filling factor of the warm ionized medium (Anantharamaiah and Bhattacharya 1986).

The work on radio recombination lines was later extended to many other contexts, including starburst galaxies and AGN (Anantharamaiah et al. 1995; Phookun et al. 1998; Mohan et al. 2002). Almost all major radio telescopes in the world were used for such studies. A survey of the galactic plane for carbon recombination lines at the low radio frequency of 34.5 MHz was undertaken some years later with the T-array at Gauribidanur (Kantharia et al. 2001).

5.5.2 The Gauribidanur T-array Radio Telescope

A large, fixed radio telescope operating at the decametric radio frequency of 34.5 MHz was set up in the early 1980s at Gauribidanur, about 80 km away from Bangalore. This was a collaborative effort of RRI with the IIA see Section 5.4.1). The telescope consisted of two fixed dipole arrays arranged in the form of a "T". The east-west arm was 1.4 km long, at the middle of which joined a southern arm of 0.45 km length. The dipoles were arranged in east-west rows with an inter-dipole spacing of 8.6 m and an inter-row spacing of 5 m in the north-south direction. The east-west arm had four rows of dipoles, each row containing 160 dipoles. The south arm had 90 rows, each containing four dipoles. Signals from these 1,000 dipoles were electrically phased and combined to form a beam of 26 arc-min × 40 arc-min at zenith (Sastry et al. 1981; Udaya Shankar and Ravi Shankar 1990).

One of the major projects undertaken with the Gauribidanur T-array (Figure 5.26) was an all-sky survey at 34.5 MHz (Dwarakanath and Udaya Shankar 1990). Data was collected in interferometric mode using a 128-channel 1-bit correlator system built at RRI, recording visibilities by correlating individual rows in the south arm with one of the rows in the east-west arm. The result of the survey provided maps of the non-thermal galactic background emission, along with other emitting and absorbing sources. Ionized hydrogen regions showed up as strong absorption features, while large non-thermal loops and spurs stood out in emission.

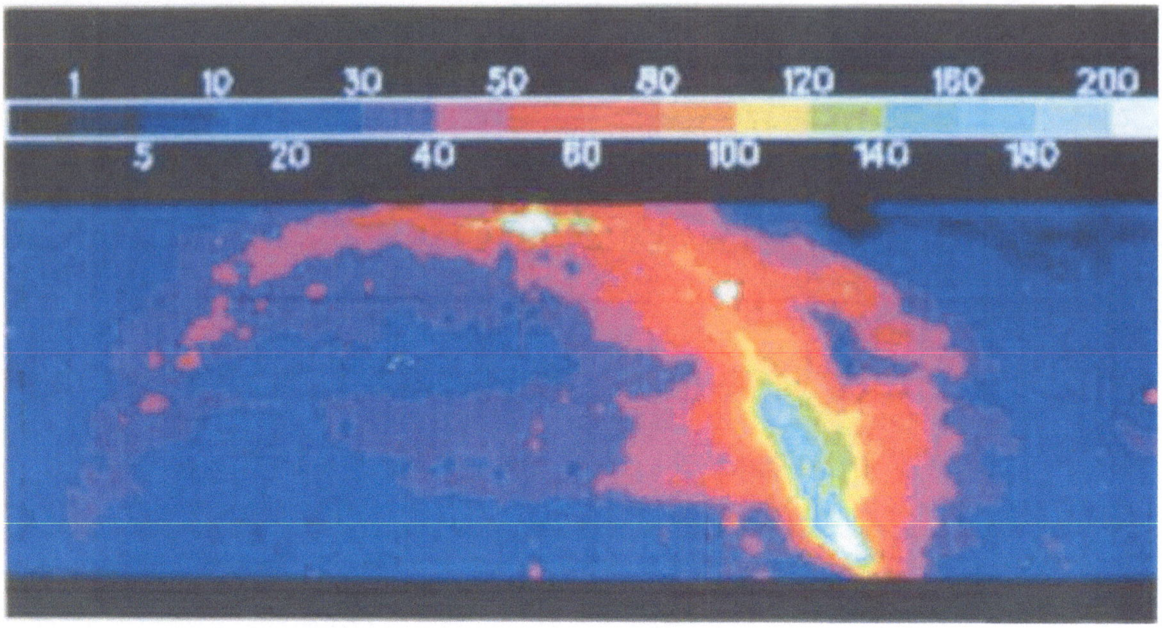

Fig. 5.26 The sky map at 34.5 MHz made using the Gauribidanur T-array. The colour key represents the scale of brightness temperature in units of 1000 K. Right ascension increases from right to left and declination from bottom (-36 deg) to top (+64 deg). Image courtesy K.S. Dwarakanath and N. Udaya Shankar

The T-array has also been used for a spectral line survey as mentioned above, and for extensive pulsar observations, which will be described in a later section.

5.5.3 The Mauritius Radio Telescope

RRI and IIA carried the experience gathered with the Gauribidanur T-array to establish a similar observatory in the southern hemisphere, on the island of Mauritius during the 1990s. In this Indo-Mauritian joint venture, a non-coplanar array of 1,024 fixed broadband helices was installed in an east-west arm of 2,048 m in length. The south arm here is 880 m long, and consists of 16 movable trolleys, with four helices on each. Visibilities are measured using a 512-channel 2-bit 3-level correlator system. At least 60 days of observations are required to cover all the spacings in the south arm up to 880 m, giving a synthesized beam size of about 5 arc-min (Golap et al. 1998). A survey of the southern sky at 150 MHz has been carried out with this telescope. The data collection was completed over several years. Preparation of maps using this data is ongoing; full resolution maps covering about 25% of the survey area, along with a source catalogue, are now available (Pandey 2006, see also http://www.rri.res.in/surveys/MRT/Download.html).

5.5.4 Instrumentation for the GMRT

The GMRT set up by NCRA-TIFR near Pune has received major contributions from the RRI. RRI undertook to equip all antennas of the GMRT with feeds and receivers covering the 21 cm hydrogen line band.

Fig. 5.27 An aerial view of the Mauritius Radio Telescope, as seen from the south. A part of the east-west arm with fixed helices is visible in the upper part of the figure. Sixteen trolleys with four helices each are seen in the south arm.

Installation of these large (~500 MHz) bandwidth, dual linear polarization front ends on the GMRT was completed by 1999. The receiver developed in this connection has also been used in a stand alone experiment to make measurements of the temperature of the Cosmic Microwave Background radiation at 1,280 MHz (Raghunathan and Subrahmanyan 2000).

Another valuable contribution of RRI to the GMRT is the user-configurable digital GMRT Array Combiner (GAC) which allows the GMRT to be operated as a single phased or incoherent array. Signal from all antennas are combined to produce a single voltage (phased array) or power (incoherent array) data stream for each of the spectral channels in each of the two polarizations (Prabu 1997). This instrument is particularly useful for pulsar studies, allowing time series analysis to be carried out on a single set of data streams rather than for each antenna separately. For pulsar studies at GMRT, RRI also built a versatile real time signal processor intended to act as a pulsar polarimeter. This processor applies on-line corrections for dispersion, Faraday rotation, Doppler acceleration and parallactic angle variations of the pulsar signal and records the stokes parameters, folding the pulses at a given period (Ramkumar and Deshpande 2002).

5.5.5 Pulsar Studies

The study of radio pulsars has been a sustained activity at RRI. Other than the GMRT, special instrumentation was also developed for the Gauribidanur T-array. Limited tracking using electrical phasing in the east-west direction was implemented; a swept-frequency local oscillator was employed for dispersion compensation (Deshpande 1992), and later a portable pulsar receiver was built and used, which recorded baseband voltages and allowed for a variety of offline processing.

A 1024-channel digital spectrometer with postdetection dedispersion facility designed and built at CSIRO, Australia (McConnell et al. 1996) was used in collaboration between RRI and CSIRO at the ORT for some

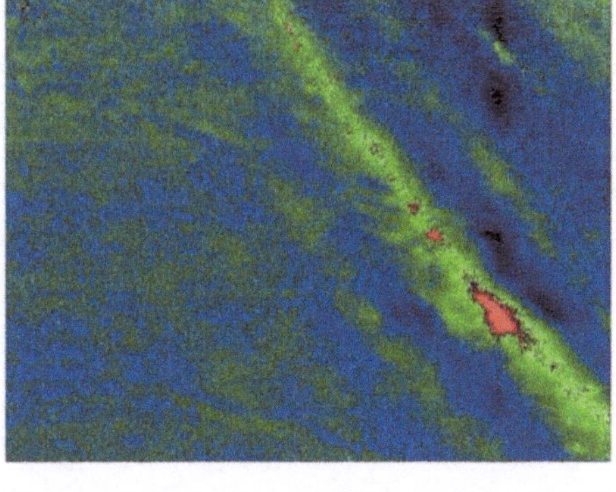

Fig. 5.28 An image of the region RA18 to 19 and Dec -40 to -5 deg at 150 MHz obtained using the Mauritus Radio Telescope (Pandey 2006)

time. This machine was later shifted to the Mauritius Radio Telescope. Among the notable results obtained during the Ooty run were the observations of intense radiation spikes from the nearby millisecond pulsar J0437-47, which were interpreted as diffraction patterns arising due to coherent emission over ~100 m lateral extent in the pulsar magnetosphere (Ables et al. 1997). Accurate measurement of scattering delay of a sample of pulsars was also carried out using this instrument (Ramachandran et al. 1997).

A cartographic transformation to map the pattern of observed subpulse drift to a rotating pattern of sparks in the pulsar magnetosphere was devised at RRI (Deshpande and Rankin 1999). While initially applied to the data obtained from the Arecibo telescope, this technique was employed also to the observations carried out using the Gauribidanur T-array (Asgekar and Deshpande 2005); (Figure 5.29)

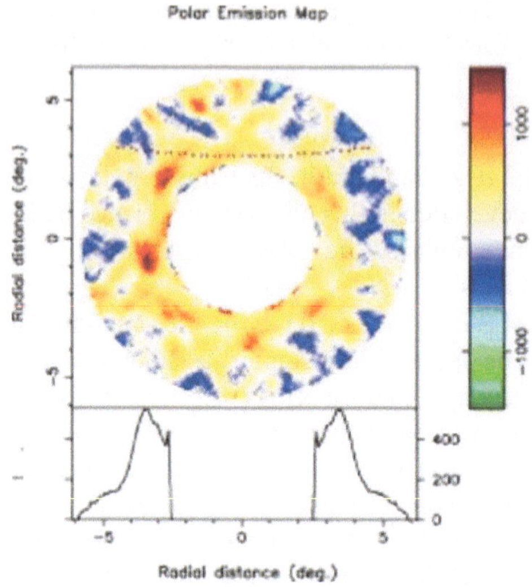

Fig. 5.29 Average polar emission map (colour coded intensity) of the pulsar PSR 0834+06 obtained via cartographic transform from the data collected using Gauribidanur T-array at 34.5 MHz. The line of dots show the traverse of the sight line through the emission beam (from Asgekar and Deshpande 2005)

The ORT continues to be in use by the RRI pulsar group. At present a long-term pulsar timing programme is ongoing.

5.5.6 Observations of Neutral Hydrogen Gas

Study of the kinematics of interstellar gas and of galaxies in clusters and groups using the 21 cm hyperfine transition of neutral hydrogen has been a sustained theme at RRI, and has been pursued using the VLA (USA), WSRT (Netherlands), ATNF (Australia) and the GMRT. Parkes telescope observations of the neutral hydrogen absorption spectrum towards the galactic centre were used to determine the velocity distribution of cold clouds. A population of high-velocity clouds was discovered this way (Radhakrishnan and Sarma 1980), and was studied later in greater detail with the GMRT and the VLA (Dwarakanath et al. 2004). An extensive study of neutral hydrogen absorption by diffuse clouds at high galactic latitude was carried out using the GMRT (Mohan et al. 2004). Large imaging surveys of neutral hydrogen emission undertaken by RRI with the GMRT include those of galaxies in the Eridanus group (Figure 5.30) (Omar and Dwarkanath 2005) and of galaxies in X-ray bright groups (Sengupta et al. 2007).

5.5.7 Millimetre Wave Astronomy

RRI initiated the effort of millimetre wave astronomy in India, starting in the early 1980s. A 10.4 m diameter millimetre wave dish antenna was indigenously built following the design of similar antennas in use at the Owens Valley Radio Observatory operated by CalTech. The dish is constructed out of individually adjustable

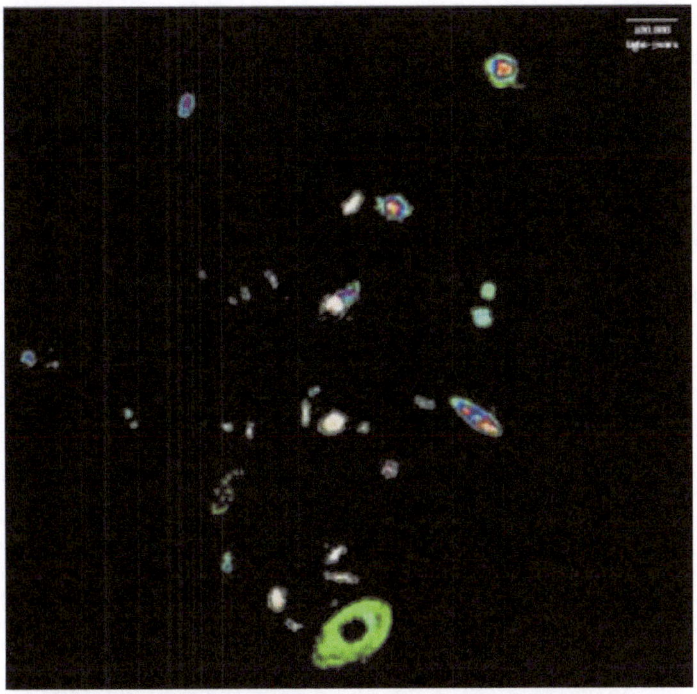

Fig. 5.30 A collage of the galaxies in the Eridanus group imaged in the 21 cm wavelength of the hyperfine transition of neutral hydrogen. The colour coding represents the neutral hydrogen column density (from Omar and Dwarakanath 2005).

hexagonal panels made of lightweight aluminium honeycomb material sandwiched between aluminium skins. The surface accuracy is 100 microns, in order to enable operation at 2.6 mm wavelength corresponding to the spectral line generated by rotational transition of the CO molecule, the primary tracer of diffuse molecular gas in the universe.

Large observational programmes carried out at the 115 GHz (2.6 mm) band with this telescope include a study of Cometary Globules in the Gum Nebula, which led to a measurement of the expansion of the system of globules (Sridharan 1992), and a survey of molecular clouds in the Gould's belt, which reconstructed a 3-dimensional structure of the distribution of molecular gas in this region (Ramesh 1994). However, due to restrictions placed by the local weather, the available time for observations at 115 GHz was quite limited. The telescope has thus spent a significant fraction of its time observing at lower frequencies. An extensive survey of SiO maser sources in the galaxy has been carried out at 86 GHz band over several years (Patel et al. 1992), and a methanol maser survey at 6.7 GHz band is currently ongoing. New receivers at 40–50 GHz band are now being installed on this telescope for a spectral line survey of star forming regions. Along with the 10.4 metre telescope, RRI also installed a smaller millimetre wave telescope made using a 1.5 metre diameter dish moulded out of synthetic material. This was used as a test bed for various technology developments before deployment at the 10.4 m dish. In addition, a study of atmospheric ozone was carried out with this telescope, deriving the height distribution of ozone over Bangalore (Vivekanand and Arora 1988).

5.5.8 Recent Developments

Radio astronomy continues to be an area of major research and development at the RRI. Some of the recent projects RRI has been engaged in are as follows:

A reconfigurable FPGA-based digital back-end has been built and installed at the ORT. This digitizes the signals from 22 modules of the telescope at the intermediate frequency (IF) stage and allows for either recording of the raw voltages or to combine them to form beams. This receiver will significantly enhance the versatility of the telescope.

RRI is currently engaged in the design and fabrication of a 40–90 MHz feed and receiver system for the GMRT. Four GMRT antennas have been equipped with the designed system and tests are in progress.

Fig. 5.31 A picture of the 10.4 m diameter millimetre wave dish located at the RRI campus. The 10.4 m dish was installed in RRI campus and has been in operation since 1988. At the highest operating frequencies, a closed cycle 4K helium cryogenic system has been used.

A wideband pulsar receiver capable of simultaneously sampling multiple frequency bands with dual polarzation is being developed. This can be used at the prime focus of any large aperture telescope, the current aim being the Green Bank Telescope in the USA.

A 12 m preloaded parabolic dish of a novel design developed and patented by Swarup and Tapde (2000) of NCRA has been constructed and installed by RRI at Gauribidanur. The characterization of the dish is currently in progress. Extensive photogrammetry measurements have been performed to provide feedback to structural analysis.

RRI has entered into a major partnership with the Murchison Wide-field Array telescope located in Western Australia. This telescope will eventually consist of 512 individual tiles of 16 dual-polarization dipoles each, optimized for operation over 80–300 MHz frequency range. Signals from these tiles will be combined at a central station. RRI has fabricated the first phase digital receiver for eight tiles, which digitizes the 16 analog signals, breaks each of them up into 256 spectral channels, selects 25 channels each (about 32 MHz bandwidth) and transmits them to the central station over fibre links.

5.6 Radio Astronomy at the Physical Research Laboratory

5.6.1 Interplanetary Scintillations, Solar Wind and Solar Studies

Introduction: The radio astronomy work was initiated at the PRL by Bhonsle and colleagues in mid 1960s who built a 10.7 cm receiver for monitoring solar activity. Later dipole arrays were built at three stations for measuring the velocity of solar wind by observing interplanetary scintillations (IPS) at 103 MHz (Alurkar et al. 1989). The arrays were used for scintillation studies of compact extragalactic sources. The three IPS arrays have been shut down since 2000 due to intense interference from nearby TV transmitters. Subsequently the radio astronomy group at PRL has been using national and international facilities.

The Three Station IPS Network at PRL: The locations of the three arrays are shown on the right in (Figure 5.32). The average baseline separation between the three telescopes was ~ 200 km, the size of the first Fresnel zone near the sun at 103 MHz. While the arrays at Rajkot (22° 18°N; 70° 44′ E) and Surat (21° 09°N; 72° 47′E) had a collecting area of 500 m^2 each, the array at Thaltej near Ahmedabad, (23° 18′; 72° 29′), had a collecting area of 20,000 m^2. The primary objectives of this network was to study the structure and distribution of plasma density inhomogeneities in the interplanetary medium between 0.3 and 1 AU, to study the angular structure of compact radio sources and to make systematic and regular measurements of solar wind velocities. (Figure 5.32) shows a portion of the dipole array at Thaltej.

Fig. 5.32 On top is seen a view of the dipole array at Thaltej, showing the reflecting plane and the individual dipoles loaded on open wire transmission lines; the inset shows the locations of the three IPS telescopes

IPS Studies of Cometary Ion Tails: The phenomenon of IPS can be used to study the plasma in the cometary ion tails along the sight lines of radio sources. Observations were carried out with the Thaltej radio telescope on two occasions, when comets Halley and Austin (1989c1) occulted compact radio sources (Alurkar et al. 1986; Janardhan et al. 1992). Estimates were made of the rms electron density fluctuations (ΔN) in the cometary ion tails.

Studies of Interstellar Scattering Using IPS: The IPS observations of extragalactic radio sources provide estimates of the angular size of their compact components. A comparison was made between IPS observations using the Thaltej array at 103 MHz and measurements made using an array at Cambridge at 151 MHz. It was found that appreciable enhanced scattering occurs in the plane of the galaxy (Janardhan and Alurkar 1993). The interstellar scatter broaening at 103 MHz was estimated to be 0.07 arc-sec for galactic latitudes $|b| > 20°$.

Ulysses Solar Corona Experiment (Solar Wind Studies Close to the sun and at High Latitudes): While the phenomenon of IPS at metre wavelengths can be exploited to study the solar wind at distances beyond approximately 40 solar radii, the plasma properties and structure of the solar wind at high latitudes and at distances <40 solar radii could be studied using dual-frequency Doppler sounding data from the Ulysses satellite's Solar Corona Experiment (SCE). The sounding data yielded solar wind velocities and measurements of columnar electron densities at southern solar latitudes between the pole and the equator in the distance range 4–40 solar radii (Janardhan et al. 1999). The Ulysses SCE was performed at the spacecraft's two solar conjunctions in summer 1991 and winter 1995.

Study of Solar Wind Disappearance Events: An extensive study of the extremely rare low-density solar wind anomalies at 1 AU, now known as "solar wind disappearance events" (e.g. Balasubramanian et al. 2003; Janardhan et al. 2008a) has shown that these events originate at the boundaries between large active regions and coronal holes located at central meridian and are caused by a process of interchange reconnection at the coronal hole and active region boundary. It has been shown that all disappearance events are characterised by extended periods of abnormally low densities at 1 AU, sometimes lower by two orders of magnitude from the average, and highly non-radial solar wind outflows. The azimuthal velocities can be as high as 100 km s^{-1}. Further, Janardhan et al. (2008b) showed that with the exception of co-rotating interacting regions, disappearance events provide the first link between the sun and space weather effects at 1 AU, arising from non-explosive solar events.

5.6.2 Solar Radio Emission and Space Weather

Space Weather: The rotation of the sun and its atmospheric disturbances are extremely complex phenomena and are likely to contribute to space weather. Vats (2007) has used 2.8 GHz solar radio emission over a period of several solar cycles to study the coronal rotation both on temporal and spatial scales. Mehta (2005) found that the coronal rotation has large temporal variations that correlate with the variability observed by optical methods, though there are differences in the radio and optical estimates of the coronal rotation. The first two components may be related to solar activity and Hale periodicities, respectively. The multi-frequency radio investigations reveal that the solar corona exhibits a differential rotation as a function of altitude (Vats et al. 2001).

5.6.3 Quasar and Pulsars

The giant quasar J1432+158 discovered using the GMRT is the largest single object known beyond a redshift of one (Singal et al. 2004). Using the Rajkot radio telescope at 103 MHz, a large number of giant pulses from the pulsar PSR 0950+08 were detected (Singal 2001). A possible detection was made of radio pulses from Geminga, a well known X-ray and gamma ray pulsar (Singal et al. 1999).

5.7 Conclusion

During the last 40 years several radio telescopes have been built in India, being amongst the best in the world. These have yielded many important results concerning a variety of radio sources in the universe. Successful design of the GMRT has played a role in the promotion of the ambitious Square Kilometre Array (SKA), as suggested by Swarup (1991) and Wilkinson (1991).

The indigenous design and construction of radio telescopes in India have also led to many spin offs. The design and construction of the ORT has contributed to growth of industries that have built a large number of antennas for satellite and microwave communication in the country. Many of the scientists and engineers who got trained in the radio astronomy groups later joined dozens of public and private institutes and industries. Four became professors in the Indian Institute of Technology at Bombay, Guwahati, Kharagpur and Madras. The Indian radio astronomy groups have also contributed significantly to the construction of radio telescopes in Brazil and Mauritius.

Important scientific contributions and discoveries have been made by Indian radio astronomers in a wide variety of topics such as radio emission from the sun, pulsars, HII regions, recombination lines, supernova remnants, centre of our galaxy, dwarf galaxies, nearby galaxies, supernovae, radio galaxies, quasars, HI studies and cosmology. The relatively lower radio noise environment in India compared to that in the western countries, has allowed construction of several large facilities for operation at metre wavelengths.

The future of radio astronomy in India seems very bright. The GMRT is being upgraded to increase its capability very significantly. RRI is also adding several new facilities. It is proposed to initiate Very Long Baseline Interferometry (VLBI) between the GMRT and radio telescopes in Australia and elsewhere. In 1985 Ananthakrishnan et al. (208, 209) carried out VLBI observations of bright radio galaxies and quasars at 327 MHZ using the Ooty Radio Telescope and several radio telescopes in USSR and Europe. VLBI Observations were done by Ananthkrishnan and Sankarraman in 2001 with few antennas of the GMRT and Russian radio telescopes. Since the GMRT has a large collecting area, I suggest that it would also be advantageous for training of students to develop an Indian VLBI network by installing modest size radio telescopes, say of 6 m diameter, at some of the Indian Institutes of Technologies (IITs), Indian Institutes of Science Education and Research (IISERs) and selected universities; this is practical today with the availability of low cost storage disks of 1,000 GB and also economical Rubidium clocks. The pressing need is to undertake novel educational training programmes which would attract more students in the field of experimental astronomy in India and its potential spin offs. There are many unsolved problems and key questions concerning the Nature that would continue to inspire mankind for decades to come.

Acknowledgements I thank Prof. T. Padmanabhan for inviting me to write this review of radio astronomy facilities and research work in India. I also thank several radio astronomers who have made valuable contributions and provided texts for different sections of this chapter. In particular, I thank P.K Manoharan for Section 5.2.3.5; R. Nityanand, J. C. Chengalur, Y. Gupta, and D. J. Saikia for Section 5.3; R. Ramesh for Section 5.4, Dipankar Bhattacharya for Section 5.5; and P. Janardhan for Section 5.6. I am indebted to Nimisha Kantharia for making a number of corrections and improvement in this review.

References

1. Ables, JG, Deshpande AA, Vivekanand M, Astrophys J (1997) **475**:L33
2. Ahuja A, Gupta Y, Mitra D, Kembhavi AK. Mon. Not. R. Astron. Soc(205) **357**:1013
3. Alurkar SK, Bhonsle RV, and Sharma AK, Nature (1986) 322:**439**
4. Alurkar SK, Sharma AK, Janardhan P, and Bhonsle RV Nature (1989) **338**:211
5. Anantharamaiah KR, J Astrophys Astr (1985a) **6**:177
6. Anantharamaiah KR. J Astrophys Astr (1985b) **6**:203
7. Anantharamaiah KR. J Astrophys Astr (1986) **7**:131
8. Anantharamaiah KR and Radhakrishnan V. Astron Astrophys (1979) **79**:L9
9. Anantharamaiah KR, Bhattacharya D. J Astrophys Astron (1986) **7**:141
10. Anantharamaiah KR, Radhakrishnan V. Astron Astrophys (1979) **79**:L9
11. Anantharamaiah KR, Zhao JH, Goss WM, Viallefond F. J Astrophys Astr (1995) **16**:273
12. Annual reports of the Kodaikanal observatory (1952) and (1962)
13. Anupama GC, Kantharia NG. Astron Astrophys (2005) **435**:167
14. Asgekar A, Deshpande AA. Mon. Not. R Astron Soc (2005) **357**:1105
15. Athreya RM, Kapahi VK (1998) **19**:63
16. Baade W, Minkowski,R Astrophys J (1954) **119**:206
17. Backer DC. Nature (1970) **228**:42
18. Bagchi J, Gopal-Krishna, Krause M, Joshi S. Astrophys J (2007) **670**:85L
19. Balasubramanian V, Janardhan P, Srinivasan S, and Ananthakrishnan S. J Geophys Res (2003) **108** A3: 1121)
20. Begum A. Chengalur J, Hopp, New Astronomy (2003) **8**:267
21. Begum A. Chengalur J, Karachentsev 2005 Astron Astrophys (2005) **433**: L1
22. Begum A. ChengalurJ, Karachentsev et al. Mon. Not. R. Astron Soc **386**, 1667, (2008), and references therein
23. Bhat NDR, Gupta Y, Rao AP. Astrophys. J (1998) **500**:252
24. Bhat NDR, Gupta Y, Kramer M, Karastergiou A, Lyne AG, Johnston S. Astron Astrophys (2007) **462**:257
25. Bhattacharyya B, Gupta Y, Gil J, Sendyk M. Mon. Not. R Astron Soc (2007) **377**:L10
26. Bird MK, Huchtmeier W, Gensheimer P, Wilson T L, Janardhan P and Lemme C. Astron Astrophys Lett (1997) **325**:L5
27. Brunetti G. et al. Nature (2008) **455**:944
28. Burke BF, Grahm-Smith F. An Introduction of Radio Astronomy, Cambridge University Press (1997) 297pp.
29. Chengalur J, KanekarN. Mon. Not. R. Astron Soc (2000) **318**:303
30. Chengalur J, Kanekar N. Mon. Not R Astron Soc (1999) (letters) 302:L29
31. Chengalur J, Kanekar Phy Rev Lett (2003) **91**:1302
32. Deshpande AA. J Astrophys Astr (1992) **13**:167
33. Deshpande AA. and Sastry Ch V. Astron Astrophys (1986) **160**:129
34. Deshpande AA, Rankin JM. Astrophys J (1999) **524**:1008
35. Doshi S, Mon. Not. R. Astro Soc (1999) **302**:L65 (1999)
36. Dwarakanath KS, Goss WM. Zhao, J.H., Lang, C.C., J. Astrophys Astron (2004) **25**:129
37. Freire PC, Gupta Y, Ransom SM. Ishwar-Chandra, Astrophys J (2004) **606**:L53
38. Gangadhara R T. et al. Astron Astrophys (1999) **342**:474
39. Gangadhara RT, Gupta Y. Astrophys J(2001) **555**:31
40. Garn T, Green DA, Riley JM. Alexander, P., Mon. Not. R. Astron Soc (2008) **383**:75
41. George SJ. Ishwara-Chandra, CH (2008) arXiv0903.2726
42. Ghosh T. Ph.D. thesis (1990), Indian Institute of Science, Bangalore
43. Ginzburg VL. Comptes Rendus de l'Academie des Sciences de l'U.R.S.S (1946) **52**:487
44. Golap K, Udaya Shankar N, Sachdev S, Dodson R, Sastry Ch V. J Astrophys Astron (1998) **19**:35
45. Gopal-Krishna. Astron Astrophys (1988) **192**:37
46. Gopal-Krishna. Astron Astrophys (1980) **86**:L3
47. Gopal-Krishna. Mon. Not. R. Astron Soc (1977) **181**:247
48. Gopal-Krishna. Patnaik AR, Steppe H. Astron Astrophys (1983) **123**:107
49. Gopal-Krishna. Preuss E, Schilizzi RT. Nature (1980) **288**:344
50. Gopal-Krishna. Steppe H, Astron Astrophys (1982)**113**:152
51. Gopal-Krishna. Swarup G, Astrophys Lett (1976)**17**:45
52. Gopal-Krishna. Swarup G, Mon. Not. R. Astron Soc (1977) **178**:265
53. Gopal-Krishna. Swarup G, Sarma NVG, Joshi MN. Nature (1972) **239**:91
54. Gupta N, Saikia DJ. Mon. Not. R. Astron Soc (2006) **370**: L80
55. Gupta N, Salter CJ, Saikia DJ, Ghosh T, Jeyakumar S. Mon. Not. R Astron Soc (2006a) **373**:972
56. Gupta N, Srianand R, Petitjean P, Khare P, Saikia DJ, York DG. Astrophys J (2006b) **654**:L111
57. Gupta Y, Gangadhara RT. Astrophys J (2003) **584**:418
58. Gupta Y, Gil J, Kijak J, Sendyk M. Astron Asrophys (2004) **426**:229

59. Gupta Y, Mitra D, Green DA, Acharyya A. Curr Sci (2005) **89**:853
60. Hey JS. Nature (1946a) **157**:47
61. Hey JS, Parsons SJ, Phillips JW. Nature (1946b) **158**:234
62. Hota A, Saikia DJ, Irwin JA. Mon. Not. R Astron Soc (2007) **380**:1009
63. Hota A, Saikia DJ. Mon. Not. R. Astron Soc (2005) **356**:998
64. Irwin JA, Saikia DJ. Mon. Not. R Astron Soc (2003) **346**:977
65. Ishwara-Chandara, Yadav JS, Rao AP. Astron Astrophys Letters (2002) **388**:L33
66. Janardhan P. J. Astrophys Astron (2006) **27**:1
67. Janardhan P, Alurkar SK. Astron Astrophys (1993) **269**:119
68. Janardhan P, Alurkar SK,Bobra AD and Slee OB. Aust J Phys (1991) **44**:5 565
69. Janardhan P, Alurkar SK, Bobra AD, Slee OB, Waldron D. Aust J Phys (1992) **45**:1 115
70. Janardhan P, Balasubramanian V, Ananthakrishnan S, Dryer M, Bhatnagar A and McIntosh PS. Solar Phys (1996) **166**, 379
71. Janardhan P, Bird MK, Edenhofer P, Plettemeier D, Wohlmuth R, Asmar SW. Patzölt, M, Karl, J Solar Phys (1999) **184**:157
72. Janardhan P, Fujiki K, Sawant HS, Kojima M, Hakamada K. and Krishnan R. J Geophys Res (2008a) **113**:A03102
73. Janardhan P, Tripathy D, Mason H. Astron Astrophys Lett (2008b) **488**:L1-L4
74. Jansky KG. Proc Inst Radio Engrs (1933) **21**:1387
75. Jennison RC, Das Gupta MK. Nature (1953) **172**:996
76. Joshi BC. et al. Advances Space Research (2006) **38**:2725
77. Joshi BC. et al. Bulletin Astron Soc India (2004) **32**:191
78. Joshi BC. et al. Bulletin Astron Soc India (2006) **34**:401
79. Joshi BC, McLaughlin MA, Kramer M, Lyne AG, Lorimer DR, Ludovici DA, Davies M, Faulkner AJ. AIPC (2008) **983**:616
80. Joshi BC, Vivekanand M. Mon. Not. Royal Astron Soc (2000) **316**:716
81. Joshi MN, Kapahi VK and Bagchi J. In Radio Continuum Processes in Clusters of Galaxies , eds. O'Dea, C.P., Uson, J. M., NRAO (1986) Charlottesville p. 73
82. Kanekar N, ChengalurJ, Astron Astrophys (2002) **382**:838
83. Kanekar N, Chengalur J Astron Astrophys (2003) **403**:43
84. Kantharia NG, Ananthakrishnan S, Nityananda R, Hota A. Astron Astrophys (2005) **435**:483
85. Kantharia NG, Anantharamaiah KR, J Astrophys Astron (2001) 22:51
86. Kantharia NG, Anupama GC, Prabhu TP, Ramya S, Bode MF, Eyres SPS, O'Brien TJ. Astrophys J (2007) **667**:L171
87. Kapahi VK. Astron Astrophys Suppl Ser (1981) **43**:38
88. Kapahi VK. Astron J (1989) **97**:1
89. Kapahi VK.: In Observational Cosmology, eds. Hewitt, A. Burbridge, G. Fang, L. Reidel, Dordrecht, p. 251 (1987)
90. Kapahi VK. J Astrophys Astron (1981b) **2**:43
91. Kapahi VK, Kulkarni VK. Astron Astrophys (1986) **165**:39
92. Kapahi VK, Kulkarni VK, Subrahmanya CR. J. Astrophys Astron (1987) **8**:33
93. Kapahi VK. Mon. Notices R. Astron Soc (1975) **172**:513
94. Kapahi VK, Saikia DJ. J Astrophys Astron (1982) **3**:465
95. Kathiravan C, Ramesh R. Astrophys J (2004) **510**:532
96. Kathiravan C, Ramesh R. Astrophys J (2005) **627**:L77
97. Kathiravan C, Ramesh R, Nataraj HS. Astrophys J (2007) **656**:L37
98. Kharb P, Shastri P. 2004, Astron Astrophys (2004) **425**:825
99. Kharb P, Shastri P. & Gabuzda DC. Astrophys J (2005) **632**:L69
100. Kochar RK, Narlikar JV. Astronomy in India. Diamond Jubilee Publication of the Indian National Science Academy, New Delhi, p. 37, (1995)
101. Konar C, Jamrozy M, Saikia DJ, Machalski J. Mon. Not. R. Astron Soc (2008) **383**:525
102. Konar C, Saikia DJ, Jamrozy M, Machalski J. Mon. Not. R. Astron Soc MNRAS (2006) **372**:693
103. Krishnamohan S, Balasubramanian V. J. Astrophys Astron (1984) **5**:483
104. Lah Chengalur, Briggs et al. Mon Not R Astron Soc (2007) **376**:L1357
105. Lal DV, Rao AP. Mon. Not. R. Astron Soc (2007) **374**:1085
106. Lal DV, Shastri P, Gabuzda DC. Astron Astrophys (2004) **425**:99
107. Loehmer O, Mitra D, Gupta Y, Kramer M, Ahuja A. Astron Astroph (2004) **425**:569
108. Lyne AG, Burgay M, Kramer MM, Possenti A, Manchester RN, Camilo F, McLaughlin MA, Lorimer DR, D'Amico N, Joshi BC, Reynolds J, Freire PCC Science (2004) **303**:1157
109. Machalski J, Kozie-Wierzbowska D, Jamrozy M, Saikia DJ. Astrophys J (2008) **679**:149
110. Manoharan PK, Ananthakrishnan S, Dryer M, Detman TR, Leinbach H, Kojima M, Watanabe T. and Kahn. J Sol Phys (1995) **156**:377
111. Manoharan PK, Ananthakrishnan S. Mon. Not. R. Astron Soc (1990) **244**:691
112. Manoharan PK, Ananthakrishnan S, Rao AP. In Proc. VI Int. Solar Wind Conf., eds. Pizzo, V.T., Sulzer, T. Sime, D.G., Boulder, Colorado, p. 5 (1987)
113. Manoharan PK. Sol Phys (2006) **235**:345
114. Manoharan PK, Tokumaru M, Pick M, Subramanian P, Ipavich FM, Schenk, K., Kaiser ML, Lepping RP, and Vourlidas A. Ap J (2001) **559**:1180
115. Martyn DF. Nature (1946) **158**:632
116. Martyn DF. Nature (1947) **159**:26
117. McConnell D, Ables JG, Bailes M, & Erickson WC. MNRAS 1996 **280**:331
118. Mehta M. Bull. Astron Soc India 2005 **33**:323
119. Menon TK. Astrophys J (1976) **204**:717

120. Mercier C, Subramanian P, Kedron A, Pick M, Ananthkrishnan S, Janardhan P. Astron Astrophys (2006) **447**:1189
121. Mitra D, Rankin JM, Gupta Y. Mon. Not. R. Astron Soc (2007) **379**:932
122. Mitra D, Rankin JM. Mon. Not. R. Astron Soc (2008) **385**:606
123. Mohan NR, Anantharamaiah KR, Goss WM. Astrophys J (2002) **574**:701
124. Mohan R, Dwarakanath KS, Srinivasan G. J Astrophys Astr (2004) **25**:143
125. Mohanty DK, Balasubramanian V. Mon. Not. R. Astron Soc (1975) **171**:17 , (Errata: Mon. Not. R. Astron Soc 1975 172:501
126. Omar A, Dwarakanath KS. J Astrophys Astr (2005) 26: 1
127. Orr MJL and Browne IWA. Mon. Not. R. Astron Soc (1982) **200**:1067
128. Pandey M. 2006. Ph. D. Thesis, Mumbai University
129. Pandey VN. 'A Southern Sky Survey at 151.5 MHz using the Mauritius Radio Telescope', Ph.D. thesis, Raman Research Institute, (2006).
130. Paredes Marti, Ishwara-Chandra, Bosch-Roman. Astrophy J Lett (2007) **654**:L135
131. Patel NA, Joseph A, Ganesan R. J Astrophys Astr (1992) **13**:241
132. Pawsey JL. Proc Inst. Electrical Eng (1950) **97**:290
133. Penzias AA, Wilson RW. Astrophys J (1965) **142**:419
134. Phookun B, Anantharamaiah KR, Goss WM. Mon. Not R Astron Soc (1998) **295**:156
135. Prabu T. Array Combiner for GMRT, M.S. Thesis (1997) Dept. of Electrical Communication
136. Prabu T. Internal Technical Report (2008) Raman Research Institute, Bangalore
137. Raghunathan A, Subrahmanyan R. J.Astrophys Astron (2000) **21**:1
138. Ramachandran R, Mitra D, Deshpande AA, McConnell D, Ables JG. Mon Not R Astron Soc (1997) **290**:260
139. Ramesh R, Ebenezer E. Astrophys J (2001) **558**:L141
140. Ramesh R, et al. Astrophys J (2006) **648**:707
141. Ramesh R, et al. Solar Phys **253**:319
142. Ramesh R, Kathiravan C, Sastry Ch V. Astrophys J (2001) **548**:L229
143. Ramesh R, Kathiravan C, Sastry. Ch. V, Astrophys J (2003) **591**:L163
144. Ramesh R, Sastry Ch. V. Astron Astrophys (2000) **358**:749
145. Ramesh R, Subramanian KR, Sastry Ch V. Astron Astrophys (1999) **139**:17
146. Ramkumar PS, Deshpande AA. J Astrophys Astr (2002) **22**:301
147. Rao AP, Ananthakrishnan S. Nature (1984) **312**:707
148. Rao AP, Subrahmanyan R. Mon Not R Astron Soc (1988) **231**:229
149. Reber G. Astrophys J (1940) **91**:621
150. Reber G. Astrophys J (1944) **100**:279
151. Roshi AR, Anantharamaiah KR. Astrophys J (2001)**557**:226
152. Roy S, Rao AP. J Phys Conf Ser (2006) **54**:156
153. Roy S, Rao AP. Mon. Not. R. Astron Soc (2004) **349**:L25
154. Saha MN. Nature (1946) **158**:549
155. Saha MN. Nature (1946) **158**:717
156. Saikia DJ, Gupta N, Konar C. Mon. Not. R. Astron Soc (2007) **375**:L31
157. Saikia DJ, Junor W, Muxlow TWB, Tzioumis AK. Nature (1989) **339**:286
158. Saikia DJ, Konar C, Kulkarni VK. Mon. Not. R. Astron Soc (2006) **366**:1391
159. Saikia DJ. Mon. Not. R Astron Soc (1984) **209**:525
160. Saikia DJ, Shastri P, Cornwell TJ, Junor W, Muxlow TWB. J Astrophys Astron (1989) **10**:203
161. Saripalli L, Gopal-Krishna, Reich W, Kuhr H. Astron Astrophys (1986)**170**:20
162. Sarma NVG, Mohanty DK. Mon. Not. R. Astron Soc (1978) **184**:181
163. Sastry Ch V, Dwarakanath KS, Shevgaonkar RK. J Astrophys Astr (1981) **2**:339
164. Sastry Ch V. Solar Phys (1973) **28**:197
165. Sastry Ch V, Subramanian KR, Krishan V. Astrophys Lett (1983) **23**:95
166. Schmidt M. Nature (1963) 197:1040
167. Selvanayagam A, Praveenkumar A, Nandgopal D, Velusamy T. IETE Technical Review (1993) **10**:333
168. Sengupta C, Balasubramanyan R, Dwarakanath KS. MNRAS (2007) **378**:137
169. ShanmughaSundaram GA, Subramanian KR. Astrophys J (2004) **605**:948
170. Singal AK. Astrophys and Sp Sci (2001) **278**:61
171. Singal AK, Konar C, Saikia DJ. Mon. Not. R Astro Soc (2004) **347**:L79
172. Singal AK, PK Manoharan PK, Richard G, Strom R.G. ASP Conf Ser IAU Symp (2002) **199**:381
173. Sinha RP, Swarup G. Proc Astron Soc of Australia (1967) **1**:41
174. Sirothia SK, Dennefeld M, Saikia DJ, Dole H, Ricquebourg F, Roland J. Mon. Not. R. Astron Soc in press (2009) arXiv0812.0813
175. Sirothia SK, Saikia DJ, Ishwara-Chandra CH, Kantharia NG. Mon. Not. R. Astron Soc (2009) **392**:1403
176. Smith FG. Nature (1951) **168**:555
177. Smits JM, Mitra D, Stappers BW, Kuijpers J, Weltevrede P, Jessner A, Gupta Y. Astron Astrophys (2007) **465**:575
178. Southworth GC. J. Franklin Institute (1945) **239**:285
179. Srianand R, Gupta N, Petitjean P, Noterdaeme P, Saikia DJ. Mon. Not. R. Astron Soc (2008) **391**:L69
180. Sridharan TK. J Astrophys Astr (1992) **13**:217
181. Subrahmanya CR. and Hunstead R W. Astron Astrophys (1986) **170**:27
182. Subrahmanya CR. Bull. Astron Soc India (1975) **3**:13
183. Subrahmanyan R. Anantharamiah, K.R., J. Astrophys Astron (1990) **11**:221
184. Subrahmanyan R, Narasimha D, Rao AP, Swarup G. Mon. Not. R. Astron Soc (1990) **246**:263
185. Subrahmanyan R,Swarup G. J Astrophys Astron (1990) **11**:237

186. Subramanian KR, Ebenezer E. Astron Astrophys (2006) **451**:683
187. Swarup G, Ananthkrishnan S, Kapahi VK, Rao AP, Subrahmanya CR, Kulkarni, VK. Current Science (1991) **60**:95
188. Swarup G. Bull. Astron Soc India (1981) **9**:269
189. Swarup G. Current Science (1991) **60**:75
190. Swarup G, Odhiambo TR, Okoye SE, Report of a Workshop held at Ootacamuna on Proposals for INISSE and Giant Equatorial Radio Telescope, Tata Inst. of Fundamental Research, Bombay 400 005, April, 1979.
191. Swarup G. J Astronomical History and Heritage (2008) **11**:194
192. Swarup G. J Astronomical History and Heritaige (2006) **9**:21
193. Swarup G, Kapahi VK, Isloor JD, Sinha RP. Nature (1966) **212**:910
194. Swarup G, Kapahi VK, Velusamy T, Ananthkrishanan S, Balasubramaian V, Gopal-Krishna, Pramesh Rao A, Subrahmanya CR, Kulkarni VK. Current Science (1991) **60**:79
195. Swarup G. Mon. Notices R. Astron Soc (1975) **172**:501
196. Swarup G. Astrophys Astron (1984) **5**:139
197. Swarup G, Sarma NVG, Joshi MN, Kapahi VK, Bagri DS, Damle SH, Ananthkrishnan S, Balasubramanian V, Bhave SS., Sinha RP. Nature Phys Sci (1971) **230**:185
198. Swarup G, Tapde, patent application India 721/MUM/2000, patent India:207809; US:6911958B2; Australia:2001278672;China:CN100416920C
199. Udaya Shankar N, Ravi Shankar TS. J Astrophys Astr (1990) 11:297
200. Vats HO. Asian J Phys (2007) **16**:291
201. Vats HO, JR. Cecatto, M. Mehta, HS. Sawant and J.A.C.F. Neri, Astrophys. J., Lett. **548**, 87-89, (2001)
202. Venturi T. et al., in Proc. of 'The Low-Frequency Radio Universe',ASP Conference Series, eds. D.J. Saikia, D.A. Green, Y. Gupta, Tizianna Venturi Vol LFRU (2009)
203. Vivekanand M, Arora RS. Current Science (1988) **57**:1103
204. Vivekanand M, Joshi BC. Astrophys J(1997) **477**:431
205. Vivekanand M. Mon. Not. Royal Astron Soc (1995) **274**:785
206. White SM, Janardhan P and Kundu MR (2000). Astrophys. J. Lett., **533** , L167
207. Wilkinson PN, In 'Radio Interoferometry: Theory, Techniques and Applications', IAU colloquium 131, ASP Conference Series (1991) eds. TJ Cornwell and RA Perley Vol **19**:428
208. Ananthkrishna S, and 11 others, Indian J. Radio and space Physics (1985) **14:17**
209. Ananthkrishna S, and 11 others, Mon. Not. R Astron Soc (1989) 237:341

Chapter 6
Growth of Optical Astronomy in India

S. S. Hasan

6.1 Historical Perspective

Until about 80 years ago, astronomy centred solely around observations in the visible region of the electromagnetic spectrum. India has a rich heritage in astronomy going back more than a millennium. The noteworthy contributions of Aryabhatta, Varahamihira, Brahmagupta, Bhaskara I and Bhaskara II are covered elsewhere in this book. However, till the time the telescope was invented 400 years ago, the human eye was the main instrument used. The use of the telescope revolutionised astronomy, signalling the birth of optical astronomy, as we know it today.

The earliest recorded use of a telescope in India was in 1651 by one Jeremiah Shakerley who observed the transit of Mercury from Surat (Kochhar and Narlikar 1995). The first observation of any scientific value was made another 38 years later by a Jesuit priest, Father Jean Richaud, who observed α Centauri from Pondicherry and discovered its binary nature (Rao, Vagiswari and Louis 1984). Then in 1792, the East India Company established an observatory at Madras "for promoting the knowledge of astronomy, geography and navigation in India". The observatory grew in the 19th century to be one of the principal institutions devoted to work on the fundamental positions of stars, and until the commencement of activity by the British at the Cape Observatory in South Africa, the Madras Observatory continued to be a principal source of stellar positions for most of the southern hemisphere stars. John Goldingham (1796–1805, 1812–1830), T. G. Taylor (1830–1848), W. S. Jacob (1849–1858) and Norman R. Pogson (1861–1891) were the successive government astronomers who directed the activities in Madras. Under Taylor's direction, in 1843, after 13 years of painstaking work with a transit instrument and a mural quadrant, a catalogue of 11,000 southern stars was produced by the Madras Observatory, eliciting praise from the then astronomer Royal Sir George Airy, who called it the "greatest catalogue of modern times".

Research in stellar and planetary astronomy continued to thrive in Madras in the latter part of the 19th century. Norman Pogson, whose name is associated with the modern definition of the magnitude scale, arrived in Madras as the government astronomer in the beginning of 1861. The observatory had recently acquired a transit circle by Troughton and Simms which was mounted and ready for use in 1862. Pogson, who had considerable experience with transit instruments in England, put this instrument to good use. With the help of his Indian assistants, Pogson measured accurately the positions of about 50,000 stars for the next 25 years. In addition to his routine work, Pogson used an eight in Cooke equatorial to observe asteroids and variable stars. He discovered several minor planets and other objects. It is appropriate at this stage to mention the contributions of an Indian astronomer C. Ragoonatha Chary, an assistant to Pogson, who in 1867 discovered the variable nature of the star R Reticuli.

Apart from his own accomplishments in astronomy, Pogson's years in Madras are remembered for another important reason – two total eclipses and one annular eclipse of the sun were visible from India during the period, to all three of which Pogson led a team from his observatory, taking an active part in the events. The first one of these, a total eclipse on August 18, 1868, created enormous interest amongst European astronomers and preparations for its observation were made in England and France for many months preceding the event. Teams of professional astronomers from both countries arrived in India and established their camps

Indian Institute of Astrophysics, Bengaluru - 560 034. e-mail: hasan@iiap.res.in at Guntoor, on the central line of the eclipse. The Madras Observatory astronomers had their camp at Masulipatam and Vunpurthy further east. This eclipse is of great historical significance as it was the first time when spectroscopes were used during an eclipse event. A new line close to the D_2 line of sodium and to the left of it was seen in the spectrum of the chromosphere.

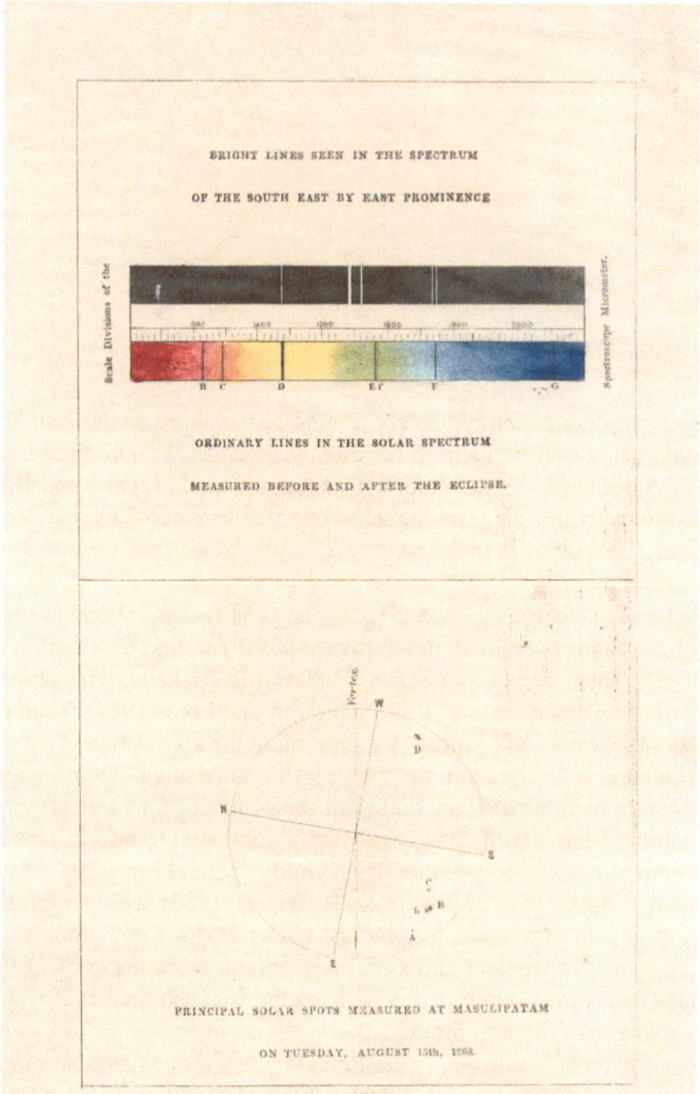

Fig. 6.1 Hand coloured sketch of the solar spectrum recorded during the total solar eclipse at Masulipatam, India on August 18, 1868 (from IIA archives)

The discrepancy in the wavelength of this line with the sodium line was confirmed by Norman Lockyer who could not ascribe it to any known terrestrial element. This new element was named helium, though it took another 27 years for its discovery by Ramsey in the laboratory. During the same eclipse, observations of the hydrogen Balmer lines in the spectra of the prominences established their gaseous nature.

Historically, the eclipse of 1868 is an important landmark associated with the birth of solar physics in India. Janssen and Lockyer made effective use of the spectrograph to show that 'red prominences could at any time be examined, without waiting for an eclipse at all'. During the annular eclipse of June 6, 1872, which was visible from Madras, Pogson found the bright chromospheric spectrum flash out for a short duration on the formation and again at the breaking up of the annulus. This is the first observation on record of viewing the flash spectrum at an annular eclipse.

In May 1882, Pogson proposed a 20 in telescope to augment the needs of his observatory. The proposal received active support both in India and Britain and the search for a suitable location in the southern highlands of India was authorized. The task was entrusted to Michie Smith, a professor of physics at Christian College, Madras, who had arrived in India in 1877. Michie Smith undertook a survey of the Palni and Nilgiri hills in 1883–1885, his observations covering the twin requirements of transparency and steadiness of image during both day and night.

In 1879, a Committee was appointed by the British Government, consisting of some leading astronomers in the country, to consider and advise on the methods of carrying on observations in solar physics. One of the tasks of the Committee, which came to be known as the Solar Physics Committee (hereafter, SPC), was to reduce the solar photographs taken daily since 1878 in Dehradun with a photoheliograph. The British Government in India had supported the work in the belief that a study of the sun would help in the prediction of the monsoons, their success and failure, the latter often leading to famines that caused such a havoc. SPC also suggested to the Government of India "that photographs of the sun should be taken frequently in order that India might assist towards securing a permanent record of the number and magnitude of the sunspots and other changes in the solar surface". At a later date they added that special spectroscopic observations should also be taken up in India, in order to collect evidence which will probably throw light on the constitution of the sun. In 1885, the Royal Society, London, constituted the Indian Observatories Committee comprised of the Astronomer Royal, and a few Fellows of the Royal and Royal Astronomical Societies.

The Committee was entrusted with the task of coordinating the work of Madras and Bombay Observatories and of advising the Secretary of State for India pertaining to the administration of these observatories. The suggestions of SPC regarding regular observations of the sun and the thinking of the Indian Observatories Committee on the future orientation of activities of the Madras Observatory converged to pave the way for the creation of a new observatory in the hills of South India. After the death of Norman Pogson on June 23, 1891, a series of initiatives were taken that eventually led to the establishment of the Solar Physics Observatory in Kodaikanal, a change in the administrative control of the observatories and the relegation of the Madras Observatory to a secondary status. It took about 9 years to effect this transition.

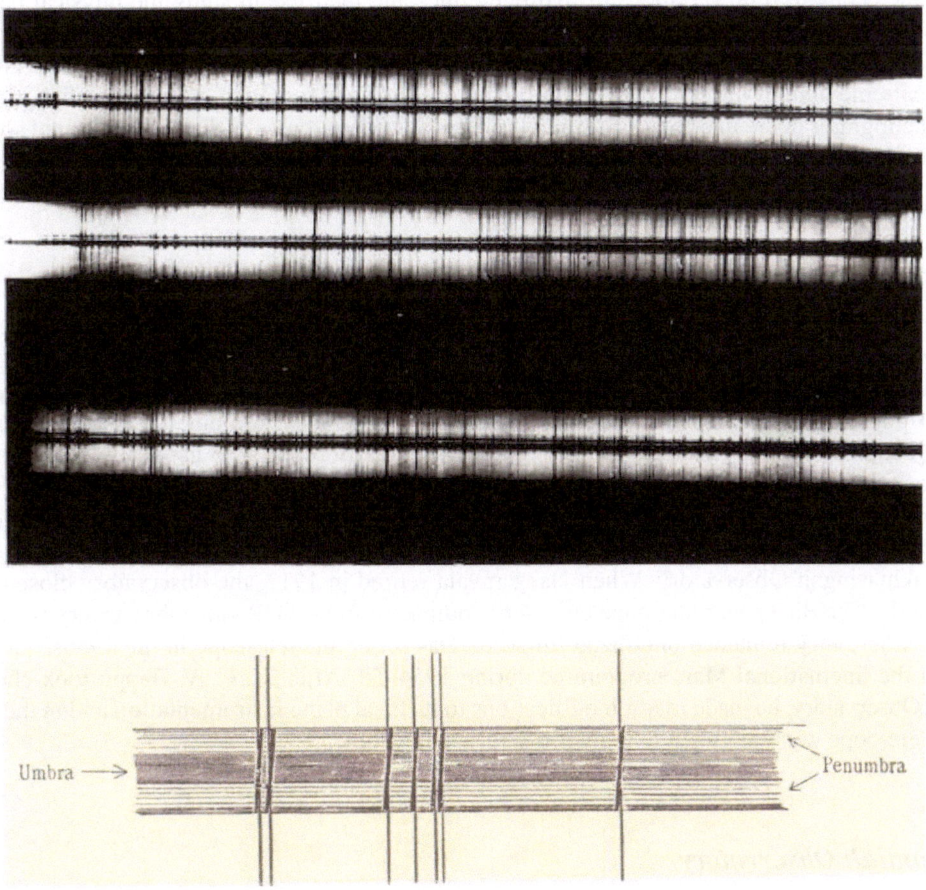

Fig. 6.2 Top: Solar spectra of a sunspot region recorded by Evershed on January 5, 1909 at Kodaikanal. Bottom: Line sketch of the spectrum showing the shift of the absorption line penumbra around the sunspot (from IIA archives)

6.1.1 Birth of the Kodaikanal Observatory

shifted to Kodaikanal. It consisted of a photoheliograph, a spectrograph, a six in Cooke equatorial, a transit telescope, chronograph, and a six in Lerebour and Secretan equatorial. Officially, the observatory started work on April 1, 1899. The principal thrust of the work at Kodaikanal was solar spectroscopy. In 1904, a Ca K spectroheliograph was acquired. This was the instrument that John Evershed put to effective use after his arrival in Kodaikanal in 1907. He also started working on an auxiliary spectroheliograph to obtain pictures of the sun in the line of H_α. Evershed's chief work in the initial years in Kodaikanal was on the spectra of sunspots. He used the high dispersion spectrographs at the observatory (there were two of them) to systematically study the spectra of spots. In the early morning of January 5, 1909, on a day when the atmosphere was exceptionally steady and the sky transparency was excellent, Evershed found two large spots and obtained their spectra. In his own words: "The spectra revealed a curious twist in the lines crossing the spots which I at once thought must indicate a rotation of gases, as required by Hale's recent discovery of strong magnetic fields in spots, but it soon turned out from spectra taken with the slit placed at different angles across a spot that the displacement of the lines, if attributed to motion, could only be due to a radial accelerating motion outward from the centre to the penumbra. Later photographs of the ionized calcium H and K lines and the H_α line revealed a contrary or inward motion at the higher levels represented by these lines." This discovery, known as the Evershed effect, was the first time that line displacements in the penumbral region were seen, indicating an outward radial flow of gases in the spots. None of the previous spectroscopic studies had revealed the dynamics of flow of the gases in the spots. It is interesting that even a century after the discovery of the Evershed effect, there is still no generally agreed consensus on its physical explanation.

Following John Evershed's retirement in 1923, activity in solar physics continued unabated at Kodaikanal and work progressed along the lines of the early years. The successive directors at the observatory were T. Royds (1923–1937), A. L. Narayan (1937–1946) and A. K. Das (1946–1960). The scientific highlights of this era were: (a) the discovery of oxygen lines in emission in the chromosphere without the aid of an eclipse, (b) the centre to limb variations of hydrogen lines and their use to study the physical nature of the solar atmosphere, and (c) the detailed study of the properties of prominences, seen in H_α. Eclipse studies constituted an important activity of the observatory, which were used for spectroscopic measurements of lines at the extreme limb of the sun. The uninterrupted observations till today of the solar disk in white light and spectroheliograms in the lines of Ca II K and H_α form a unique archive at the observatory of solar activity for over a century.

6.1.2 Takhtasinghji Observatory and the Bhavnagar Telescope

In 1888, an observatory was established in Pune through a donation by Maharaja Takhtasinghji of Bhavnagar and a matching grant from the Government of India. The observatory was headed by K. D. Naegamvala, a professor of physics at Elphinstone College in Bombay, who had an enormous interest in astronomy. The observatory acquired a 20 in Grubb reflector telescope (popularly referred to as the Bhavnagar telescope), the largest one in the country at that time, that was used for both photometry and spectroscopy. It also had several smaller instruments, which were used mainly for eclipse observations. The most notable of these were the observations of the solar corona in 1898 and are described in volume I of the publications of the Maharaja Takhtasinghji Observatory. When Naegamvala retired in 1912, the observatory closed down due to lack of funds. The Bhavnagar telescope and other equipment from the Poona Observatory were shifted to Kodaikanal, where they remained unutilized till A. K. Das set up the telescope in the 1950s. The telescope was used in the international Mars programme during 1954–55. After M. K. V. Bappu took charge of the Kodaikanal Observatory, he made major modifications to its focal plane instrumentation and in the 1960s, the Bhavnagar telescope was finally put to regular use for stellar studies.

6.1.3 Nizamiah Observatory

In 1901 a rich nobleman of Hyderabad, Nawab Zafar Jung, started a private observatory at Begumpet (Hyderabad) with a 15 in refractor and an eight in astrograph. It was formally established as the Nizamiah

Observatory in 1908 by a *firman* of the 6th Nizam of Hyderabad, Nawab Mir Mahboob Ali Khan Bahadur. The following year, the observatory participated in an ambitious international programme "Carte-du-Ciel" or astrographic sky survey. Its aim was to map the whole sky photographically by assigning various celestial zones covering declinations from 17–23 degrees south and 36–39 degrees north. This effort resulted in the publication of 12 volumes of astrographic star catalogues comprising observations of around 800,000 stars. The study of comets, variable stars, lunar occultations and solar activity was also pursued at the observatory. The availability of a spectroheliоscope in the 1930s and a blinker comparator extended the sphere of activity of the observatory. Proper motions studies of stars in the Hyderabad astrographic zone commenced and from 1944, studies of double stars were taken up. The observatory also participated in solar activity studies during the International Geophysical Year (1957–58) and the International Quiet sun Year (1964–65). Observations of lunar occultations, variable stars, proper motion of star clusters were some of the other programmes that were carried out.

6.1.4 Other Observatories

One of the earliest modern observatories, is the one in Calcutta that was established by the East India Company in 1825 mainly for routine observations, and also for studying lunar transits. The main instruments were a transit telescope, an alt-azimuth circle and a telescope.

In 1832, the Royal Observatory of Lucknow was set up by Nawab Nasiruddin Haider of Oudh. Equipped with a mural circle, a transit telescope, an equatorial telescope and astronomical clocks, it was used for studying planets, asteroids and lunar occultations. The observatory was closed down in 1849 after the death of its director Wilcox. It was destroyed during the 1857 war.

Mention should also be made of an observatory founded in 1837 by the Maharaja of Travancore at Travancore (now Thiruvananthapuram). It was equipped with a transit instrument, two mural circles, an equatorial telescope and instruments for magnetic and meteorological studies. Another observatory that was established in 1878 was at Dehradun. Set up at the suggestion of Norman Lockyer, it obtained daily photographs of the sun using equipment procured in 1874 to observe the transit of Venus. Daily photography of the sun continued in Dehradun till 1925.

In Calcutta, two observatories were established at St. Xavier's College in 1875 and at the Presidency College in 1900. The former had equatorial telescopes, transit instruments and spectroscopes. Solar prominences were regularly observed in the initial years, though subsequently it was used for meteorological and teaching purposes. The observatory at Presidency College was equipped with a 4.5 in Grubb reflecting telescope which was purchased from a donation by the Maharaja of Tripura. In 1922, it received an eight in telescope as a gift from the Astronomical Society of India, which was started in 1910 at the time of appearance of Comet Halley.

6.1.5 Post-war Development of Astronomy

In 1945, the Government of India appointed a committee with Meghnad Saha in chair to plan the post-war development of astronomy in India. The Committee had several sittings – some in Kodaikanal, some in Bangalore. In its final recommendation, the Committee suggested a long-range plan and some immediate initiatives. The latter contained a modernisation and expansion plan for the Kodaikanal Observatory and the establishment of small observatories at the universities of Delhi, Aligarh and Banaras (now Varanasi) to be used for teaching and training graduate students. The long-range plan was to establish a stellar observatory somewhere in northern India equipped with a large telescope and other auxiliary equipment. Three years later a Standing Advisory Board was created to implement the plans for expansion of the Kodaikanal Observatory. A survey was initiated to look for a suitable site in the Himalayas for a high altitude observatory. The director of the Kodaikanal Observatory, A. K. Das, also visited the plains of central India – Ujjain and Aurangabad, in order to prospect for possible sites for a central stellar observatory. As we shall see later, it took several decades before a proper high altitude observatory could be set up.

Fig. 6.3 The solar tower telescope at Kodaikanal

6.2 The Indian Institute of Astrophysics

6.2.1 Kodaikanal Observatory

In 1960, M. K. V. Bappu took over as the Director of the Kodaikanal Observatory and in the next couple of years the newly installed solar instruments came into regular use, especially the solar tower telescope. By the end of 1965 a high dispersion spectrograph and a Babcock type magnetograph were operational, which were used primarily for measuring magnetic and velocity fields on the sun.

6.2.1.1 Instruments

The solar tower telescope is used for high spatial and spectral resolution work. It consists of a 60 cm aperture two mirror coelostat mounted on a 11 m high tower platform that directs the sun's rays via a flat mirror into an underground tunnel to an achromat that forms a 34 cm diameter solar image at the focal plane. The main areas of investigation include high resolution studies of: (a) the solar chromosphere, (b) the Evershed effect, and (c) 5 min. oscillations in the solar photosphere. A significant contribution using this instrument was the identification that bright fine mottling in the chromosphere is responsible for the Wilson-Bappu effect that relates the Ca II K line-widths to the luminosity of stars.

A Littrow-type spectrograph and a spectroheliograph are the main focal plane instruments. Recently an in-house built dual beam spectropolarimeter was installed as a back-end instrument to the Littrow spectrograph on the tower telescope that is used mainly for studying sunspots.

6.2.1.2 Current Research

The current programmes include measurement of magnetic fields at different heights in the solar atmosphere, using Kodaikanal white light images to determine the solar diameter and its variation with the solar cycle, dy-

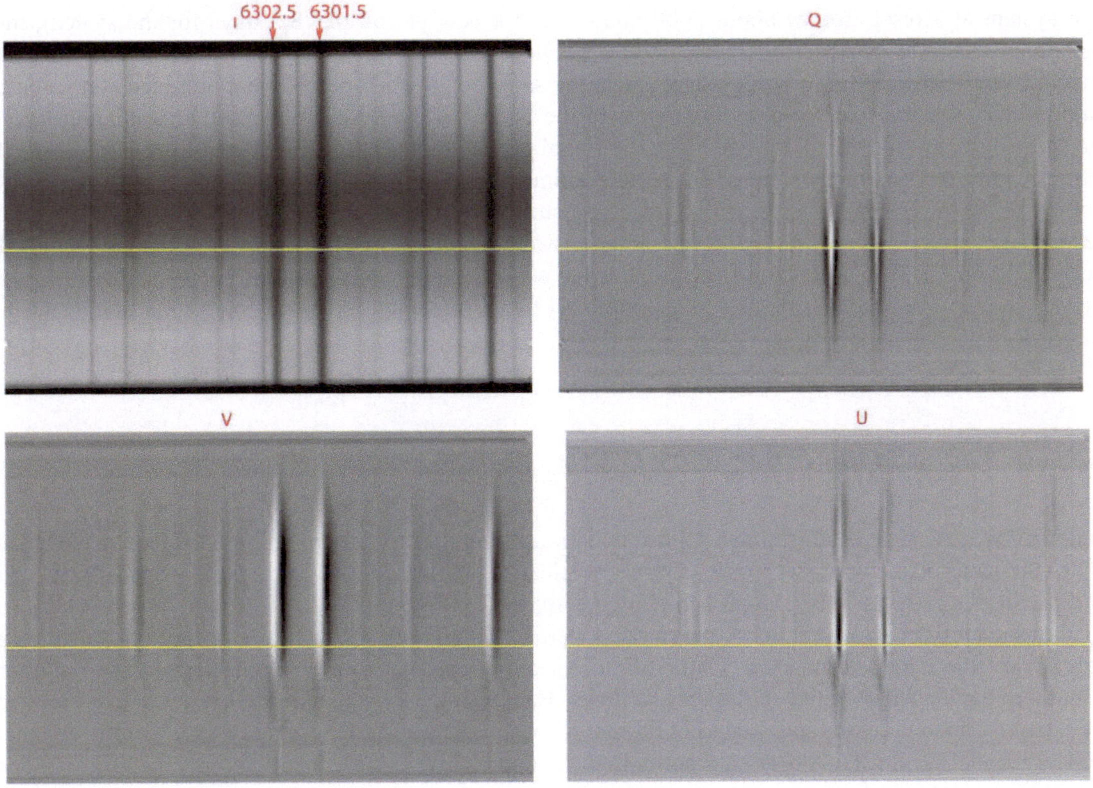

Fig. 6.4 Spectral images of the stokes parameters observed at the umbral region of the sunspot NOAA 0743 using the dual beam spectropolarimeter

namics of sunspots, solar irradiance variability, solar rotation and synoptic studies. In addition, investigations of explosive events, activity and solar cycle effects are also being carried out.

The magnetic structuring from the photosphere to the chromosphere above several active regions has been examined. The spatial variation of the chromospheric magnetic field suggests that these concentrated horizontal fields are the foot points of omega-shaped magnetic structures which are conspicuous in the X-ray images of the same active region.

Recently, an ambitious programme has commenced at Kodaikanal to digitize 100 years of solar data consisting of white light images (since 1904) and spectroheliograms in the Ca II K and H_α lines. The aim is to study long-term variations of solar features (such as sunspots, plages and filaments), irradiance, differential rotation rate and so on and their correlation with the solar cycle. The digitized data will produce the first truly comprehensive long-term data set that will be available to the astronomy community.

6.2.2 Bhavnagar Telescope

To augment the capabilities of the Kodaikanal Observatory in stellar astronomy, Bappu initiated efforts to install the Bhavnagar telescope in a proper building with a motorised moving dome and manually operated shutters. He changed the original Newtonian mount of the telescope to a Cassegrain system. He also decided to build a suitable low resolution spectrograph for it. The spectrograph came to be known as the Bhavnagar spectrograph, although it had nothing to do with either the Maharaja of Bhavnagar or Naegamvala's Observatory in Poona. The Bhavnagar spectrograph had refracting optics although the dispersive element was a reflection grating. This spectrograph was used by Bappu and all subsequent astronomers for at least 20 or more years - first in Kodaikanal and later at the new observatory in Kavalur.

In Kodaikanal, Bappu and Ganesh used the spectrograph to determine the period of the well known Wolf-Rayet binary system Gamma Velorum (Ganesh and Bappu 1967). The primary is a W-R star brighter than the secondary, a B2 subgiant, by about 2.5 magnitudes. Bappu and Ganesh obtained more than a hundred spectra of the system at a resolution of about 2000 and derived a new period of 78.5 days for the system, much longer than what was believed till then, the two previous determinations having yielded 16 and 24 days. Somewhat later Nova Delphini was observed with the same set up and spectra were obtained at a resolution of about 5000. A survey of the early type stars in the Scorpio-Centaurus association was initiated and the same spectrograph was used to determine the rotational period of the B type members of the association. The limiting magnitude was about V≈6. This work continued later in Kavalur. But Bappu's principal interest, the study of stellar chromospheres, could not be pursued with this moderate setup. A bigger telescope and a better spectrograph with a much higher resolution were needed for that. Bappu with his colleagues thus concentrated on Ca II K line studies in the sun for a very detailed investigation of solar chromospheric activity. The Bhavnagar telescope was transported to Leh in the 1980s for surveying high altitude sites near Leh.

6.2.3 The Vainu Bappu Observatory (VBO)

Bappu's mind was set on establishing a solid foundation for optical astronomy in India. He knew Kodaikanal was not the ideal place for stellar work. Cloud-free clear nights were few. Developing proper infrastructure was difficult. As early as 1962 Bappu had started a search for a good site in the southern peninsula for an optical observatory. At the time, the southern skies were explored only with a few telescopes in Australia and South Africa; the large telescopes in Chile and in Australia were yet to be commissioned. The richness of the southern Milky-Way offered endless possibilities. Bappu's plan was to acquire first a 1 m class telescope to get started on stellar work and then to build indigenously a larger 2 m class telescope, thereby enhancing the observational facilities in stellar astronomy in the country as well as augmenting our capabilities in optics and precision mechanical engineering.

Fig. 6.5 Top: The 2.34 m VBT at the VBO, Kavalur. Below: A closer, full view of the telescope inside the dome

In a short span of 10 years Bappu was able to achieve this goal. Kavalur was found to be the ideal site for the new observatory. Mid-way between Bangalore and Madras, the location was in a dense forest and protected and far from any significant light pollution. At a latitude of 10° N, the site permitted coverage of a significant part of the southern sky. The large magellanic cloud (LMC) was just about at the limit of its reach. The 'seeing' was found to be reasonably good. Operations started with a 38 cm telescope in a roll-off shutter in 1968, but Bappu had already obtained approval of the government to buy a 1 m telescope and it was ordered from Carl Zeiss, Jena. The telescope arrived in Kavalur in 1969 and started operating in January 1972. In 1971, the observatory was converted in to a fully autonomous institution called the Indian Institute of Astrophysics.

For spectroscopy, the Bhavnagar spectrograph was the main instrument used with the newly acquired 1 m Zeiss telescope. Survey of the rotation periods of the early-type stars in Scorpio-Centau association continued with it. One could now reach a V magnitude of about 8.0. The detector was still a photographic plate and various measures like baking and pre-flashing were taken to improve its efficiency so that fainter magnitudes could be reached. The 1 m telescope was equipped with a variety of focal plane instruments. These included cameras for fast photography, photoelectric photometers, a single-channel photoelectric spectrum scanner, a quartz-prism calibration spectrograph and a conventional coudé spectrograph with two different cameras, a Schmidt system, called coudé A, with a focal length of 61 cm to yield a dispersion of 12.8 Å/mm in the blue (Ca II K line) and a long 285 cm focal-length camera, called coudé B, yielding a dispersion of 2.8 Å/mm in the blue. With photographic plates, the coudé B system could only reach a limiting magnitude of 3.0. This conventional spectrograph fell into disuse after 3 or 4 years. Meanwhile, a coudé echelle spectrograph was also setup in a short time to facilitate the completion of a thesis project. A resolution between 30,000 and 50,000 was obtained with this instrument, depending upon the wavelength region observed. Image intensifiers were used regularly in conjunction with photographic films to go fainter and to achieve a better signal-to-noise ratios with this instrument. Eventually, photographic films were replaced by CCD chips as detectors and stars brighter than about the 5th magnitude were observed with ease at resolutions between 15,000 and 20,000. A substantial amount of work came out with the use of this instrument. Later an InSb infrared photometer, a Universal Astronomical Grating Spectrograph (UAGS) from Zeiss and a polarimeter were added to the complement of instruments at the Cassegrain focus.

6.2.3.1 Vainu Bappu Telescope

Soon after the institute was formally established, Bappu revived the idea of indigenously building a 2 m class telescope. The first proposal was made in 1971 and as its scientific justification, three major areas of study were identified: (i) the spiral structure of the galaxy, particularly in the southern hemisphere, (ii) probing stellar chromospheres through high-resolution spectroscopy, and (iii) morphological aspects of external galaxies and their chemical composition parameters for elucidation of aspects of stellar evolution in large stellar aggregates. A concept report was prepared in 1976, the same year when the headquarters of the institute shifted from Kodaikanal to Bangalore. A low-expansion glass blank had already been imported from Carl Zeiss, Jena and the work on shaping and figuring it started at the newly built Optics Workshop in Bangalore. An expert Committee recommended that the telescope be located in Kavalur.

Work on the 2.34 m telescope began in 1976 and the installation of the telescope was finally completed in 1985. Bappu did not live to see the successful completion of his dream project. He succumbed to a heart condition in August 1982. On January 6, 1986, the telescope was formally dedicated to the nation by the then Prime Minister of India, the late Shri Rajiv Gandhi. On that historic occasion, the Prime Minister named the telescope and the Kavalur Observatory after their creator M. K. Vainu Bappu. The Vainu Bappu telescope (VBT) has an f/3.25 prime focus, an f/13 Cassegrain focus, and a CCD-imaging camera at the prime with a 3-element Wynne corrector. It has a high-resolution echelle spectrometer which is directly fed by an optical fibre from the prime focus. Other medium resolution spectrographs are attached to the Cassegrain focus. These instruments allow wide-field prime focus imaging and medium to high resolution spectroscopy. A medium resolution spectropolarimeter is also available at the Cassegrain focus which has been used for cometary studies. With the addition of the fibre-fed coudé echelle spectrometer, it is now possible to obtain spectra of stars of the 9th or the 10th magnitude with a resolution of 30,000–100,000.

6.2.3.2 Scientific Highlights

Solar System Astronomy

A significant contribution from the VBO came from the successful observation, in June, 1972, of the occultation of the star SAO 186800 by Ganymede, indicating the presence of a thin atmosphere around this Jovian satellite (Carlson et al. 1973). The programme of occultation observations from the VBO continued and yielded important results. In March 1977, while making photometric observations of the planet Uranus during a stellar occultation event, the astronomers in Kavalur stumbled upon the discovery of the Uranian ring system (Bhattacharyya and Kuppuswamy 1977, Bhattacharyya and Bappu 1977). Evidence for the suspected existence of an outer ring around Saturn was obtained in 1984. Mutual eclipses of Galilean satellites involving Io and Europa were recorded. These were used to evaluate the tidally induced declaration of Io's mean motion (Vasundhara 1994). Mutual events of the Pluto-Charon system and lunar occultations of stars in optical and near-infrared for high angular resolution measurements have been carried out.

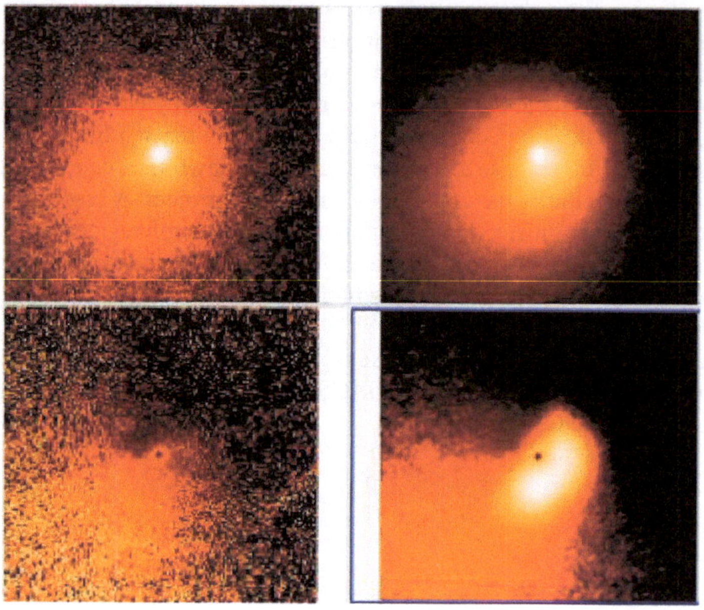

Fig. 6.6 Deep impact with comet 9P/Temple 1

In January 1987, Indian Institute of Astrophysics (IIA) launched Project Kalki to survey and discover asteroids, comets and the elusive tenth planet of the solar system. A 45 cm Schmidt telescope at the VBO was used for the survey. It had a field of view of $3° \times 4°$ with a plate scale of 150 arc sec/mm. During the few years that Project Kalki was in operation, it discovered a total of six asteroids. On February 17, 1988, the first discovery using this facility was reported. The new asteroid, 1988 DQ1, was named Ramanujan, after Srinivasa Ramanujan. Of the remaining five, two more asteroids, viz. asteroid 7564 (1988 CA) and asteroid 8348 (1988 BX) have recently been named Gokumenon and Bhattacharyya, respectively after M. G. K. Menon and J. C. Bhattacharyya.

Comet Halley was observed and studied in detail in 1985–1986. Accurate astrometric positions were estimated; fine structural details in the outer coma, hourly changes in the ionic tail, and evolution of emission from the cometary atmosphere were monitored. Properties of dust in the comet were studied using occultation of stars by the comet's coma. Imaging of the nucleus of the comet Swift-Tutle (1992) at the VBO showed that it had a rotation period of about 3 days. Comet Hale-Bopp was studied in detail (Vasundhara and Chakraborty 1999). Multiple fragments from the crash of comet Shoemaker-Levy 9 into Jupiter was observed during 16–22 July, 1994 from the VBO. An intense infrared flash in the H-band caused by the impact of fragment S was detected on 21 July 1994, using the 75 cm aperture telescope at the VBO (Bhatt 1994). On 4 July 2005, many observatories around the world and in space observed the collision of deep impact with comet 9P/Temple 1 and its aftermath. This was an unprecedented coordinated observational campaign. The observations show that (i) there was new material after impact that was compositionally different from that seen before impact;

(ii) the ratio of dust mass to gas mass in the ejecta was much larger than before impact; (iii) the new activity did not last more than a few days, and by 9 July the comet's behaviour was indistinguishable from its pre-impact behaviour; and (iv) there were interesting transient phenomena that may be correlated with cratering physics. The VBO participated in this campaign and monitored the event with CCD imaging (Meech et al. 2005). Some other comets studied are C/2004 Q2, 81P/ Wild 2 (Vasundhara and Chakraborty 2004).

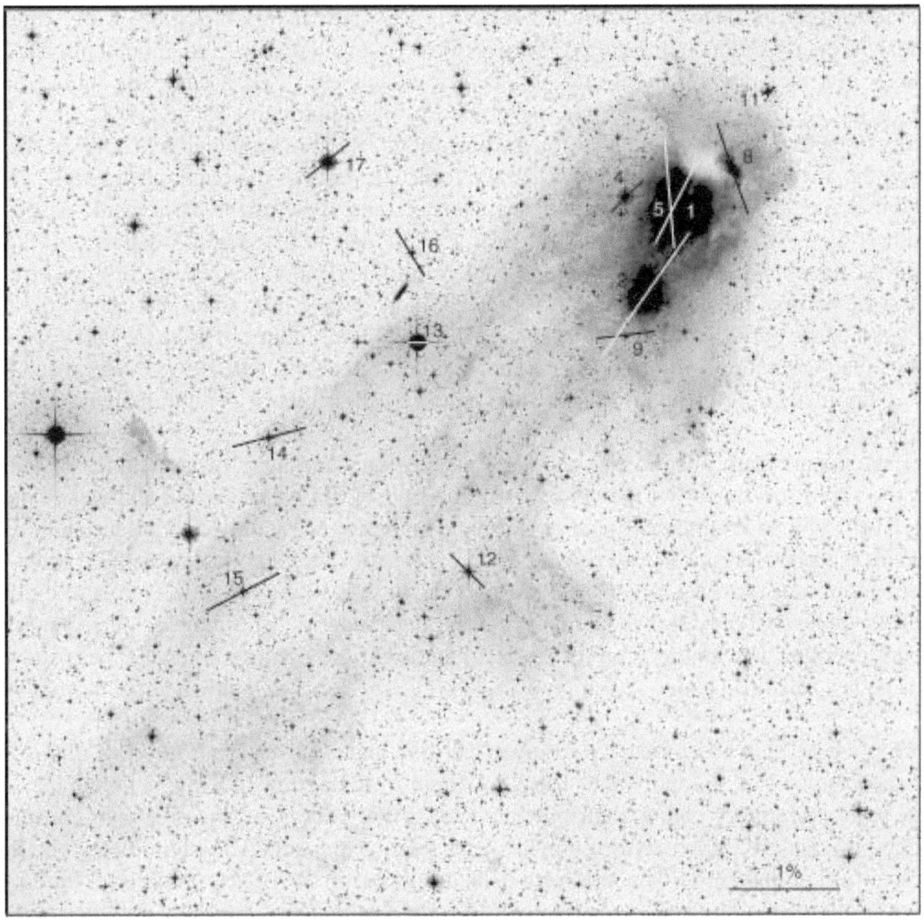

Fig. 6.7 Polarization map of the cometary globule CG12

Stellar and Galactic Astronomy

A variety of observational programmes involving spectroscopy, photoelectric and imaging photometry and polarimetry have been pursued at the VBO using the available telescopes ranging in size from 38 cm to 2.34 m in diameter.

A detailed spectroscopic study of the Scorpio-Centaurus association was completed during the early years of operation of the 1 m telescope (Rajamohan 1976). Included in the study were observations of high resolution helium line profiles of hot stars in the association, which directly provided a vindication of the then new non-local thermodynamic equilibrium models of hot stellar atmospheres. Since then many more studies on stellar clusters of different types, star-forming clouds, dense cometary globules and stellar associations have been done yielding a wealth of new results. A study of a few southern globular clusters, including 47 Tuc, ω Centauri, was conducted involving equi-densitometry of the clusters on photographs obtained in different wavelength bands. For ω Centauri, a more detailed study was done, where photoelectric scans were obtained with different apertures along the major and minor axes of the cluster and the data were used to study the ellipticity of the stellar distribution in the cluster from the centre outward (Scaria 1979). Distribution of the blue stars in ω Centauri was found to be elliptical, in contrast to the red star distribution, which appeared spherical (Scaria and Bappu 1981).

Interstellar clouds including dense cometary globules were studied polarimetrically to map the geometry of the magnetic field in these objects (Bhatt et al. 2004). It was found that magnetic fields played an important role in controlling the morphology and the process of star formation in the clouds. A number of T Tauri stars belonging to the Taurus association, Herbig Ae/Be stars and stars in clusters were monitored for photometric, spectroscopic and polarimetric variability to understand the evolution of their circumstellar disks (Mekkaden 1999, Manoj et al. 2006). The isolated T Tauri star TW Hya was studied in detail (Mekkaden 1998). Recently, the VBO participated in an international campaign, when this star was simultaneously observed in the optical and in X-rays for 2 weeks to study the relationship between its optical variability and X-ray emission.

Observations of open clusters in the galaxy were analysed to evaluate the fundamental physical parameters of these objects, e.g. their masses, ages and heliocentric distances. In each case, the initial mass function (IMF) of the cluster was derived to determine the slope (Subramaniam and Sagar 1999), which is useful for testing theories of stellar evolution and also for studies of the structure of the galaxy. One of the first papers to appear using the VBT observations, was a CCD-photometric study of the open cluster NGC 2818, perhaps the only known cluster in the galaxy to have a planetary nebula (PN) of the same name associated with it (Surendiranath et al. 1990). A limiting magnitude of V = 21 was reached and a distance of 3.8 ± 0.1 kpc was estimated for the cluster and the associated PN. Evolutionary models were used to derive a lower limit of 2.58 $M_\odot$ for the mass of the progenitor star of the nebula. A deeper photometric study on the cluster NGC 2453, a few years later, definitively proved that the PN NGC 2452, which was suspected until then to be a member of the cluster, is actually a foreground object (Mallik, Sagar and Pati 1995). Several candidate binary open clusters were imaged and studied to study the nature of their binarity (Subramaniam and Sagar 1999; Anupama et al. 1994).

Individual stars, stars with peculiar properties, variable stars and stars at late stages of their evolution have been studied spectroscopically for elucidation of their physical properties and analysis of their chemical composition. The bright southern supergiant α Carinae (Canopus) was studied in great detail by Bappu and his colleagues in the early days at Kavalur. A reported observation of an emission reversal in its Ca II K line had spurred the interest of the astronomers at IIA. However, high-resolution coudé spectra of the star obtained at the 1 m telescope never really showed a clear reversal over a period of a number of years. After Bappu's death, his younger colleagues wrapped up the study, obtaining more spectra. A few of these showed a hint of the reversal. The last scientific paper by Bappu describing the results was published posthumously (Bappu et al. 1984).

Another programme that was pursued with great perseverance and yielding good results is the photometry and H_α spectroscopy of spotted stars - the RS Canum Venaticorum (RS CVn) variables. These binary systems show light variations which are attributed to spot activity on the surfaces of their chromospherically active members, the scale of activity being much larger than on the sun. The photometry obtained over the years has helped in building up a database on a selected group of RS CVn variables. The data have been subjected to extensive modelling and used effectively to interpret many of the characteristics of these systems (Parthasarathy et al. 1981; Raveendran et al. 1982; Mohin and Raveendran 1987, 1989, 1992, 1993a, 1993b, 1994; Raveendran and Mohin 1995). The VBO participated in the international Multi-Site Continuous Spectroscopy (MUSICOS) campaign in 1992 to observe the Herbig Ae star AB Aur for studying its chromosphere and the azimuthal structure of its wind. The observatory also took part in the whole earth telescope or WET observations to study stellar pulsations in AM CVn, G 29-38, PG 1159-035, GD 358, V471 Tau, BG CMi, and RXJ 2117+34 (Solheim et al. 1998).

The study of hydrogen-deficient stars belonging to the class of R CrB type variables has been a major programme at VBO. Photometry and polarimetry of a number of these stars were carried out. Several R CrB type variables were followed in their deep minima and also during the recovery phase. With the installation of the coudé echelle spectrometer at the VBT, it has recently been possible to do high-resolution spectroscopic studies of these objects. The data are used to analyse and understand the circumstellar environment of these stars and the episodic production and ejection of dust from their outermost layers. Broad-band multi-wavelength polarimetric observations of several carbon-rich RV Tauri stars were obtained at the VBO. The observed amplitude of linear polarization of 14% in the U band in AR Pup is the highest polarization seen in any late-type star not associated with a nebulosity (Raveendran 1999). Polarimetric studies were also done of post-asymptotic giant branch stars with circumstellar dust shells (Parthasarathy et al. 2000).

Several classical and recurrent novae have been spectroscopically monitored from the VBO during the outburst and a few during quiescence. Spectroscopic differences and similarities between individual novae in outburst were studied and the physical parameters of the ejected shell estimated. Evolution of their photospheric radii and temperature were also monitored in some cases. Accretion disk spectra were obtained from observations of novae in quiescence and mass transfer rates and geometrical parameters of these discs were derived. Spectroscopic monitoring of T CrB over a long baseline in time has shown secular as well as orbital phase dependent variation in the strengths of the emission lines. Using the images of the shell of GK Per, the proper motion of the individual knots was measured and the velocity deceleration derived (Anupama and Prabhu 1991, 1993; Hric et al. 1998).

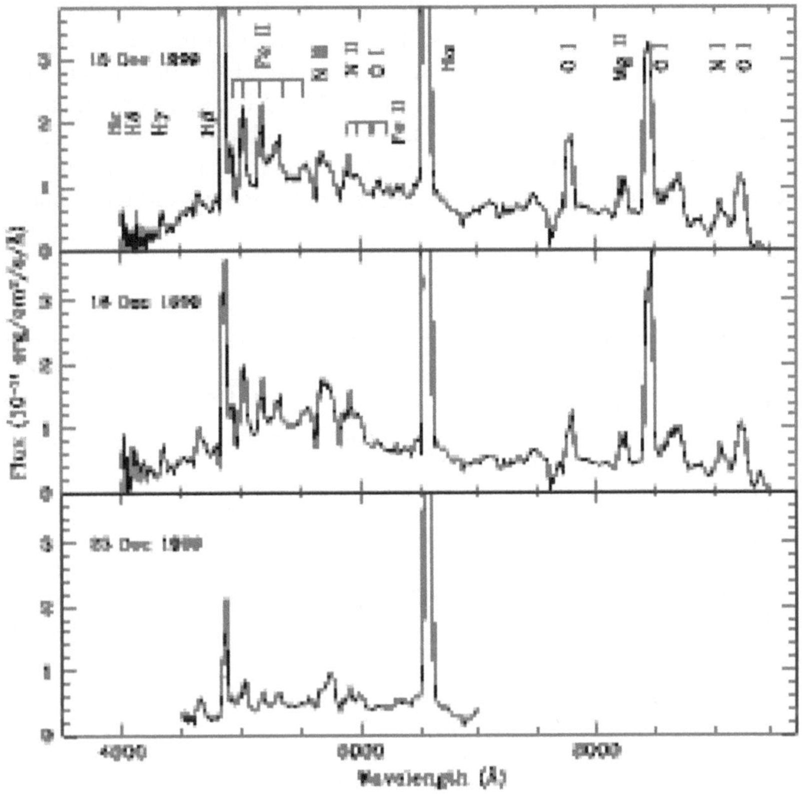

Fig. 6.8 Spectra of Nova V1494 Aql in the early decline phase

The coudé echelle spectrograph attached to the Zeiss telescope reached only about the fifth magnitude in red with CCD detectors. Until the advent of the fibre-fed echelle spectrometer at the VBT, this was the only high-resolution instrument available at the VBO. The lack of a high resolution instrument to look at fainter stars forced the observers to use the spectrosocopic facilities at other major observatories of the world where such facilities exist. Thus even before the new echelle spectrometer at the VBT was commissioned, a large body of high quality work was done by the astronomers at IIA and a number of fruitful international collaborations had emerged. A comprehensive abundance analysis of the warm galactic R CrB variables came out of observations obtained at CTIO, Chile and the McDonald Observatory in Texas, USA (Lambert and Rao 1994, 1996). An analysis of the chemical composition of the carbon-rich RV Tauri star IW Car showed that its photospheric abundances are similar to the depleted abundances observed in one of the interstellar clouds in the line of sight to ζ Ophiuchi (Giridhar, Rao and Lambert 1994).

An early piece of work with the coudé echelle spectrograph at the 1 m telescope was a study of H_α in the chromosphere of late G and early K supergiants. The blue asymmetry of the H_α profile was attributed to the occurrence of chromospheric expansion of these stars eventually leading to mass loss (Mallik 1982). The photographic study was later superseded by CCD spectroscopy. The observations were matched with elaborate transfer calculations to obtain the rates of mass loss in these stars (Mallik 1993). Another early study was concerned with the chemical composition of classical Cepheids and the related chemical inhomogeneities of the galactic disc. The [Fe/H] index of nearly two dozen Cepheids was determined. The techniques of synthetic spectra were used to obtain the chemical composition of the stars (Giridhar 1983). A survey of the infrared Ca II triplet lines was carried out with the same spectrograph in a large sample of dwarfs, giants and supergiants (Mallik 1994). The triplet lines are a potentially powerful tool to study the late-type stellar populations in galaxies and the high resolution work done at the VBO complemented the work done elsewhere at lower resolutions enabling a more accurate analysis of galaxy populations. Be stars and Be X-ray binaries were also observed at high resolution to monitor the rapid variability in their H_{α} line profile. Continuous monitoring of a few of these stars was undertaken to study the nature of the secondary and its contribution to the circumstellar disk.

The light element lithium with its principal resonance line at 670.7 nm is a major diagnostic tool to probe the physical and evolutionary processes in stars and the early history of the galaxy. The study of lithium in stars of various types thus received a great deal of attention at the VBO. High resolution studies were done of the element in a large sample of late type stars. The relationship between lithium abundance and rotation was explored in a sample of F and G dwarfs and subgiants (Mallik, Parthasarathy and Pati 2003). More recently, a search for lithium rich K giants and emission line stars in young clusters has been initiated using low resolution spectra. Candidates identified in the survey are being observed using high resolution spectra for detailed studies.

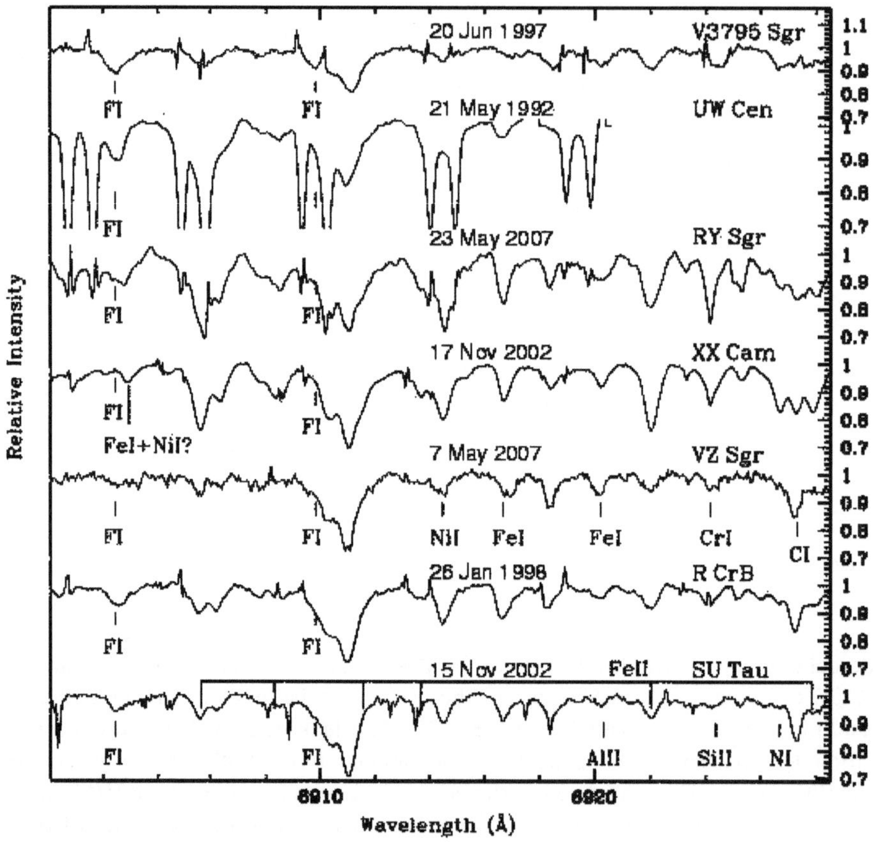

Fig. 6.9 Detection of neutral fluorine lines in cool EHe stars

High-resolution spectra of R CrB type variables at maximum light and of cool extreme helium (EHe) stars obtained at the VBO and the McDonald Observatory in Texas, USA have revealed for the first time considerable enrichment of fluorine in these objects. The abundance of fluorine in R CrB type and cool EHe stars is about 1,000 times the solar abundance (Pandey et al. 2004; Pandey 2006). This has far-reaching implications for the nucleosynthetic processes at crucial evolutionary stages of these stars. The observed chemical composition and the overabundance of fluorine and neon in these stars seem to support a double-degenerate scenario for the formation of the sequence of hydrogen-deficient stars from hydrogen-deficient carbon stars (HdCs) to R CrB to EHe stars.

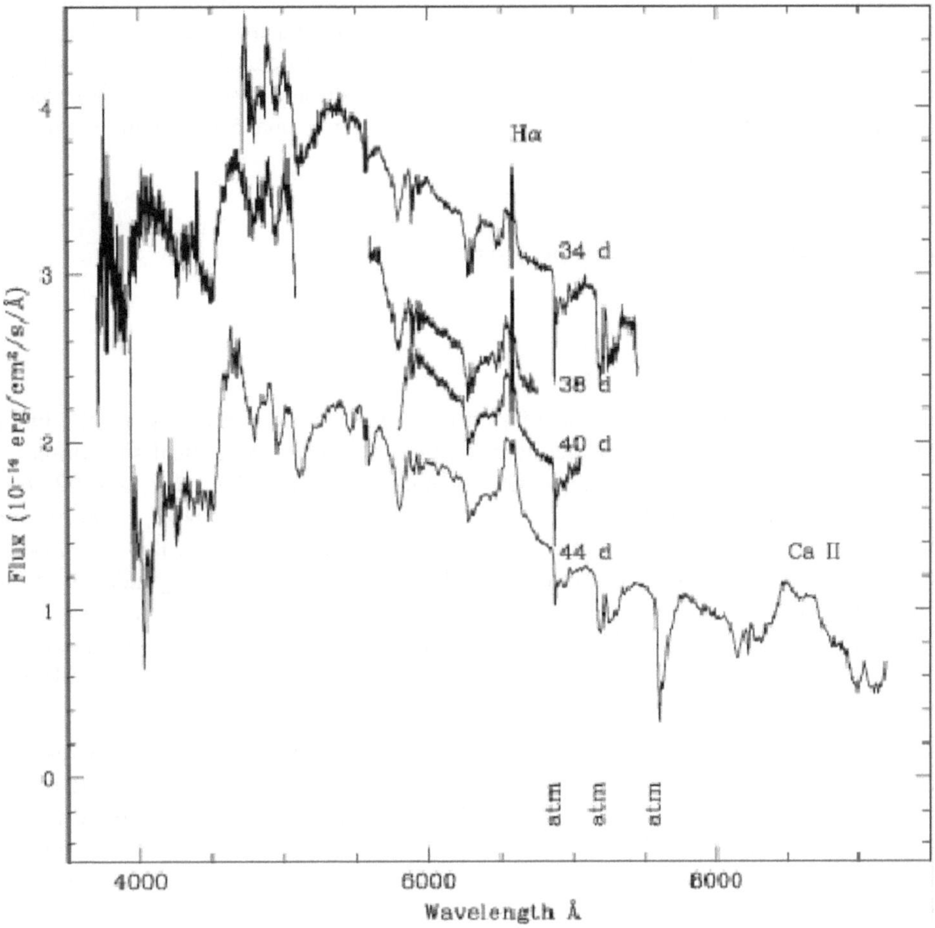

Fig. 6.10 Spectra of supernova 1998s

Extragalactic Astronomy

Soon after the commissioning of the 1 m telescope in Kavalur, Bappu initiated a survey of red stars in the direction of the LMC using ultra-low resolution objective prism spectra. A majority of the stars were found to be M giants and supergiants, and carbon stars, belonging to the LMC. A photographic study of 50 Sersic-Pastoriza galaxies was carried out a couple of years later as part of a doctoral thesis. A classification scheme was suggested that reflects the intensity and evolutionary stage of the nuclear star burst regions. The brightness of the nuclear and circumnuclear components in the barred galaxies in the sample was correlated with the length of the bar indicating the role of the bar in the supply of gas to the nucleus (Prabhu 1980).

Bulk of the extragalactic work at the VBO was done after CCD imaging became available in the late 1980s, at the prime focus of the VBT and at the Cassegrain focus of the 1 m telescope. An extensive photometric study of 161 H II regions in 9 galaxies was carried out to study star formation in galaxies (Mayya 1995). The data were interpreted with the help of evolutionary population synthesis models. Imaging of V, R and H_α in several early type galaxies, selected primarily on the basis of their X-ray emission, was used to measure nebular emission from their central regions. The presence of ionized gas, dust rings and lanes indicated that the galaxies had accreted material. The VBO participated in the international active galactic nuclei (AGN) watch by monitoring spectroscopically and photometrically the nucleus of the Seyfert galaxy NGC 3783 (Stirpe et al. 1994). The combined data showed considerable variability of the optical continuum and H_β emission with light curves resembling the ultraviolet light curve while the near-infrared light curve of the nucleus appeared independent. Photometric observations were carried out to detect intra-night variability in radio-quiet QSOs (quasi-stellar objects) in order to constrain theoretical mechanisms for their microvariability (Gopal-Krishna et al. 2000). Gamma-ray burst sources have also been studied more recently and observations from the VBO have been combined with data obtained from other observatories to elucidate the nature of these objects (Cowsik et al. 2001).

About two dozen supernovae have been observed spectroscopically at VBO, near their light maxima. The supernova types and expansion velocities of the ejecta were estimated from the data. A few of these, SN 1987A in LMC and SN 1993J in M 81, were monitored for longer periods. Observations of supernovae from the VBO have provided evidence for nitrogen enrichment in the surface layers, implying processing via the CNO-cycle in the pre-supernova progenitor (Prabhu et al. 1995; Anupama, Sivarani and Pandey 2001).

6.2.4 The Indian Astronomical Observatory

6.2.4.1 Genesis

In an attempt to establish a large optical/near infrared telescope facility in the country, the Indian Institute of Astrophysics undertook the exercise of identifying a high altitude astronomical site. Based on a survey of the available topographical maps, weather data and satellite imagery, done during 1992–93, six potential sites were identified in the Himalayan and trans-Himalayan regions, all above 4,000 m. Simultaneous reconnaissance survey of all the six sites indicated the trans-Himalayan sites, located in the rain shadow of the Himalayan mountains, to be more favourable. Of these, Digpa-ratsa Ri (longitude 78° 57′ 51″ E and latitude 32° 46′ 46″ N), Hanle was chosen for further evaluation. A continuous monitoring of the weather and cloud coverage, which began in 1995, together with occasional seeing measurements, indicated Hanle to be a world-class astronomical site for the national large optical telescope. Following the identification and characterization of the site, the necessary infrastructure was put in place.

The highest peak in Digpa-ratsa Ri is at an altitude of 4,517 meters. An area of 600 acres, including the mountain top and the plains at its base constitute the Indian Astronomical Observatory. Site characterization since 1995 has proved this location to be a stable and dark location. On an average, around 190 photometric and 260 spectroscopic nights are available for observations. The site has an average night sky brightness of 21.3 mag/arc sec^2, with low extinction (V = 0.12 mag) (Parihar et al. 2003; Stalin et al. 2008). The median seeing at the site is estimated to be 0.9–1″.

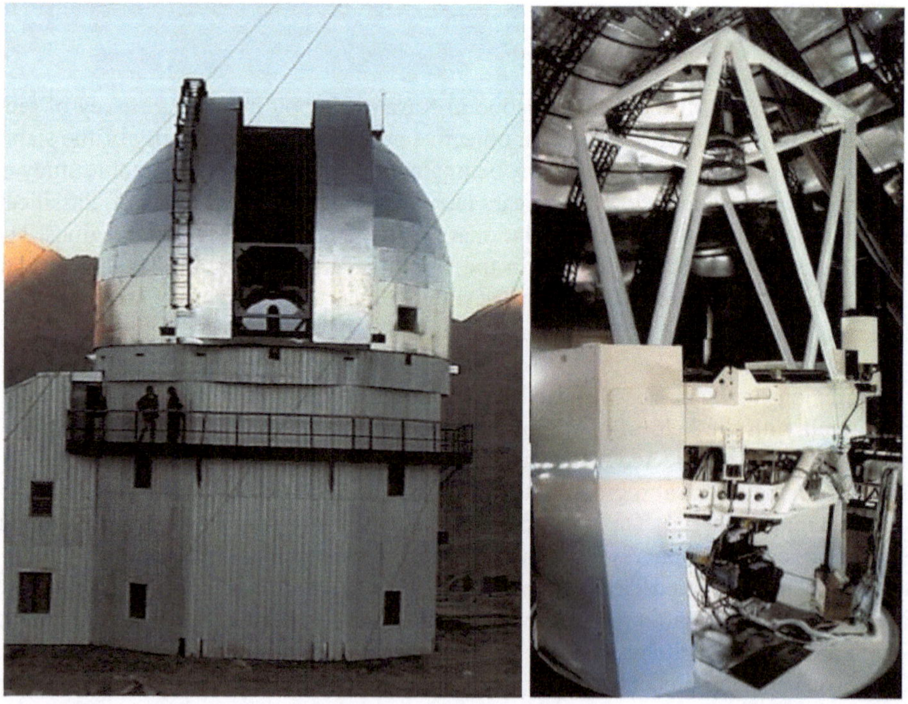

Fig. 6.11 Left: The 2 m HCT dome, with the telescope seen through the dome slit. Right: A closer, full view of the telescope inside the dome

6.2.4.2 Himalayan Chandra Telescope

The 2 m Himalayan Chandra Telescope (HCT) is situated at the mountain top, at an altitude of 4,500 m above mean sea level. It was dedicated to the nation in August 2001 after the telescope and dome were fully automated and remotely operated using a dedicated satellite communication link. It was equipped with state-of-the-art instruments for optical and near-infrared imaging, and optical spectroscopy.

The HCT is an alt-azimuth telescope with Ritchey-Chrétien design. It has an f/1.75 primary and an infrared-optimized secondary. The Cassegrain focal ratio is f/9 providing an image scale of 11 arc sec/mm. The secondary focus and tip-tilt are computer controlled to keep the optical alignment fixed at all orientations and temperatures. Pointing model corrections made online provide a blind pointing accuracy of 2.5 arc sec (rms) and good open-loop tracking. An autoguider provides accurate tracking of the telescope for long integrations. The telescope and instruments are controlled remotely via a dedicated satellite link from the institute campus at Hosakote, about 35 km from Bangalore.

The main focal instrument, mounted on the axial port, is the Himalaya Faint Object Spectrograph Camera (HFOSC). Built by the Copenhagen University Observatory, it is used for optical imaging and spectroscopy. An optical CCD imager is mounted on one of the side ports. In addition, there is a near infrared camera, mounted on another side port, which can take broad and narrow band images in the 1–2.5 micron waveband.

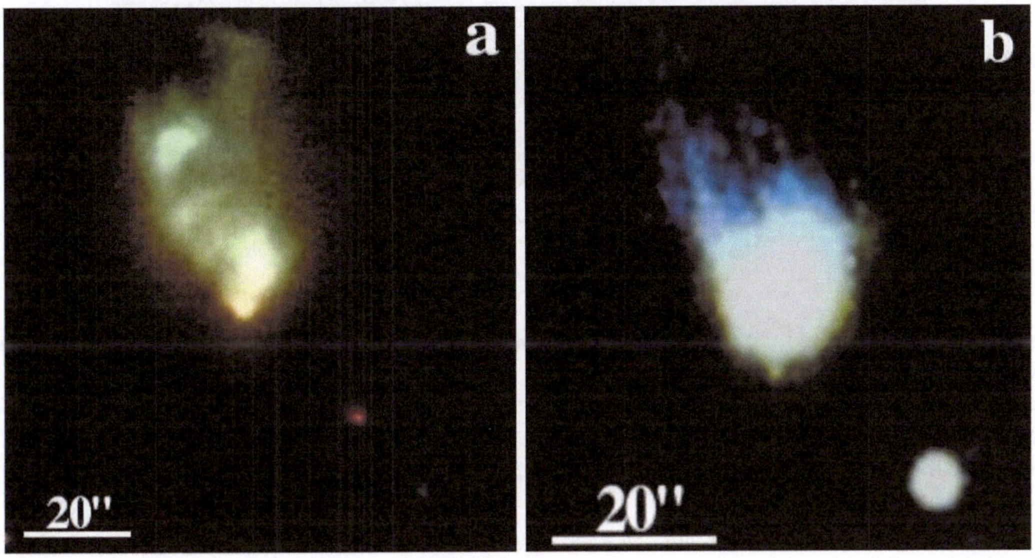

Fig. 6.12 NIR (left) and optical (right) images of the Mc Neil's nebula

6.2.4.3 Science with the HCT

Star Formation Processes and Young Stellar Objects

The HCT is being used to study galactic star forming regions, as well as regions of star formation in external galaxies to understand the details of this complex process (e.g. Pandey et al. 2008; Bhavya et al. 2007). A sample of bright rimmed clouds (BRC) surrounding such regions is being studied to quantitatively testify the "small-scale sequential star formation" hypothesis around these regions (Ogura et al. 2007; Chauhan et al. 2009). Quantitative age gradients have been found in all the BRCs studied. There is evidence that a series of radiation-driven implosion processes proceeded in the past from near the central O star(s) towards the peripheries of the H II regions.

A long-term programme to monitor young stellar objects (YSOs) in the orion nebula cluster was initiated in 2004, with an aim to address various phenomena associated with young stars. The prime motivations of this project are (a) to explore various manifestations of stellar magnetic activity in very young, low-mass stars, (b) to search for new pre-main sequence eclipsing binaries, and (c) to look for EX Or and FU Or like

Fig. 6.13 BV, H_α composite image of the infrared bright galaxy NGC 1084

transient activities associated with YSOs. Several new variables have been detected in the region, and this work clearly demonstrates the need for a systematic, long-term monitoring to detect variablility in YSOs (Parihar et al. 2009). A detailed multi-coloured monitoring of the interesting object V982 Ori over several years has established this object as a UX Or type, a rare class YSO. A detailed study of the post-outburst phase of McNeil's nebula (V1647 Orionis) (Ojha et al. 2006), confirmed that V1647 Ori is a pre-main-sequence star of the EX Or type.

A study of star forming regions in blue compact dwarf galaxies Mkn 104 and I Zw 97 (Ramya et al. 2009) indicates neither of these galaxies is a young system; instead they are undergoing episodic star formation

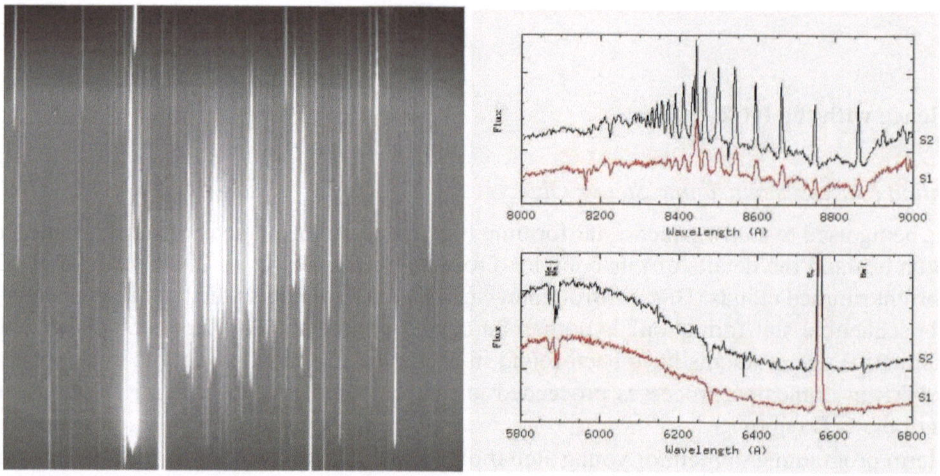

Fig. 6.14 Left: Slitless spectra of stars in the open cluster NGC 7419. Right: Spectra of three emission line stars in an open cluster, identified using slitless spectra

superposed on a faint older component. Both galaxies are very similar in their stellar content, showing an older 4-Gyr population, an intermediate 500-Myr population and a more recent burst of star formation of age 5–13 Myr. A study of the infrared bright galaxy NGC 1084 indicates that star formation in NGC 1084 has taken place in a series of short bursts over the last 40 Myr or so (Ramya et al. 2007).

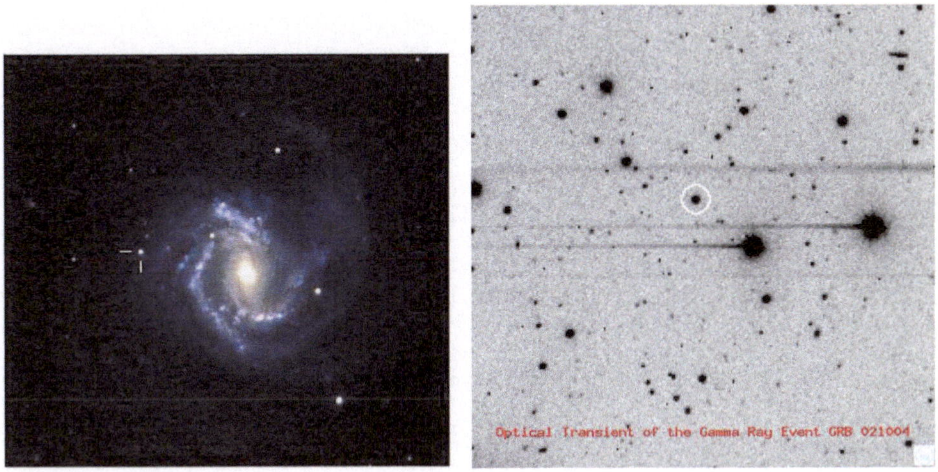

Fig. 6.15 Left: BVR composite image of the supernova SN 2008 (marked by a semi-cross) in NGC 4303. Right: R band image of the optical afterglow of the GRB 021004 (circled)

6.2.4.4 Young Open Clusters

Several groups are studying open clusters with a view to obtain accurate photometry of clusters that have so far not been sufficiently well investigated (Sujatha and Babu 2006; Sujatha et al. 2006; Carraro et al. 2006; Sharma et al. 2007).

Variable stars are being identified and followed up in order to understand the evolutionary stage of clusters. Subramaniam et al. (2005) studied the young open cluster NGC 146 and discovered several pre-main sequence stars and one Herbig Be star, while NGC 7419 was found to be a young open cluster with a number of very young intermediate mass pre-main sequence stars (Subramaniam et al. 2006; Mathew et al. 2008) performed a survey of 207 young open star clusters to identify emission-line stars using slitless spectroscopy. They identified 157 emission stars in 42 clusters and found 54 new ones in 24 of them, out of which in 19 clusters emission-line stars were found for the first time. They have also studied the Be phenomenon in these clusters.

Evolved Stars

Stars that are somewhat more massive than the sun often show peculiarities as they evolve beyond the main sequence through a dredge up mechanism that can mix the matter enriched by nucleosynthesis in the inner regions into the outer envelope. Carbon-rich stars resulting from such mixing are studied by Goswami et al. (2006). A survey of high galactic latitude CH stars (Goswami 2005; Goswami et al. 2007) and lithium rich stars has been undertaken.

The Hanle telescope is also used to study stellar oscillations in RR Lyrae type stars (Ferro et al. 2004; Ferro et al. 2008), white dwarfs (Szkody et al. 2007), and sub-luminous B type stars. Other areas of investigation include the variability in brown dwarfs, which are the missing link between stars and gas-rich giant planets like Jupiter (Maiti et al. 2005; Maiti 2007).

Stellar Explosions

It is well known that stellar explosions such as supernovae and gamma-ray burst (GRB) sources are caused by the death of massive stars and evolution of the explosion and of the remnant are determined by parameters such as the mass, metallicity and environment of the progenitor star. The high luminosities of these objects enable them to be observed at cosmological distances and make them excellent probes to study the universe at various redshifts. Supernovae of type Ia have been traditionally used as cosmological standard candles. This

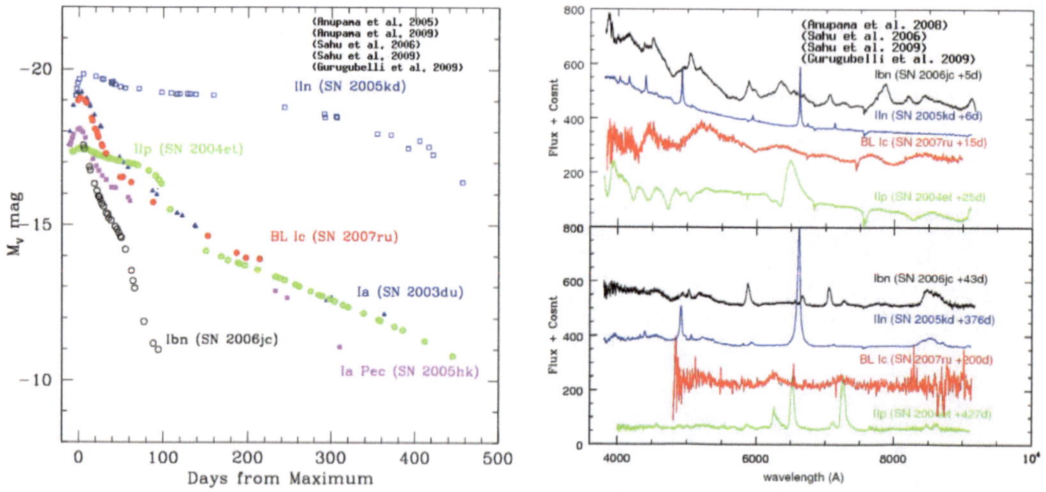

Fig. 6.16 Evolution of luminosity of supernovae of different types (left) and spectrum (right) of core collapse supernovae of different types observed with the HCT

requires good calibrations of their peak magnitudes, which can be obtained only through a detailed study from the early to the late phases of the outburst. An important goal of studying core-collapse supernovae (CCSNe) is to deepen our understanding of the progenitors and the explosion mechanisms of CCSNe. The diversity of supernovae also excites an interest in the study of the phenomenon itself and the nature of the progenitors. With these motivating factors, low redshift supernovae are being monitored with the HCT as a target of opportunity programme.

The first objects to be observed during the science verification phase of the HCT were the type Ic SN 2002ap (Pandey et al. 2003a), and type Ia SN 2002hu (Sahu et al. 2006) and SN 2003du (Anupama et al. 2005a). Several other supernovae have been observed and studied in detail since then. Some of the most interesting studies have been those of the type Ib SN 2005bf (Anupama et al. 2005b; Tominaga et al. 2005), which showed the presence of a thin hydrogen envelope, the underluminous, peculiar type Ia SN 2005hk (Sahu et al. 2008), the highly reddened type Ia SN 2003hx (Misra et al. 2008), the type Ibn SN 2006jc (Anupama et al. 2009) a type Ib supernova with narrow He emission lines, and the broad line type Ic SN 2007ru that showed a high kinetic energy to ejected mass ratio (Sahu et al. 2009).

The optical afterglows of several GRBs have been observed with the HCT. The afterglows monitored in detail include GRB 010222 (Cowsik et al. 2001), GRB 021004 (Pandey et al. 2003b), GRB 021211 (Pandey et al. 2003c), GRB 030226 (Pandey et al. 2004) and GRB 030329 (Resmi et al. 2005).

Novae are stellar explosions of a more modest magnitude in low mass binary systems. The HCT has been used to study the outburst of several classical novae and the recent 2006 outburst of the recurrent nova RS Ophiuchi (Anupama et al. 2008). An optical and radio study of the nebular remnant of the classical nova GK Persei (Anupama and Kantharia 2005) indicated the remnant to be very similar to supernova remnants, in particular, Cas A. A faint bipolar nebula, probably associated with an older planetary nebula that is associated with GK Per was detected in the lines of [N II] and H_{α}. Banerjee and Ashok (2004) used the HCT to study the nova-like variable V4332 Sagittarii and found that it did not conform to the known class of novae. Galactic microquasars and X-ray binaries (Kaur et al. 2008) are the other kinds of interactive binaries that are studied with the HCT.

Galaxies and Cosmology

The HCT is being used for several studies in the area of galaxies and cosmology. Notable amongst these studies are the study of dust formation in early type galaxies (Patil et al. 2007), the study of Wolf-Rayet galaxies and low surface brightness galaxies (Das et al. 2007).

The study of variability over various timescales in the AGN forms another important component of extra-galactic astronomy pursued with the HCT (Raiteri et al. 2005; Goyal et al. 2007).

The telescope has been used for deep J band imaging of high redshift QSOs (Goto and Ojha 2006) and is also used for multi-site monitoring of selected gravitationally lensed objects with a view to estimating the Hubble constant at all epochs.

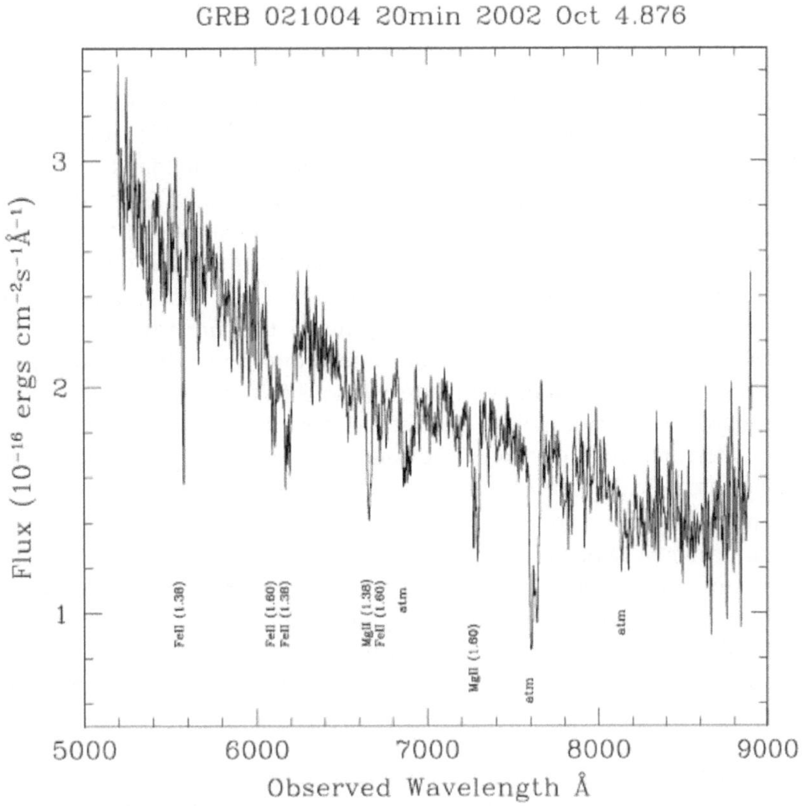

Fig. 6.17 Optical spectrum of the afterglow of GRB021004

Solar system objects

The HCT is also used for studies of solar system objects such as asteroids, comets, clouds in the atmosphere of Venus, and space debris in our immediate environment. Images of comet Temple were obtained before and after the deep impact collision as a part of an international campaign (Meech et al. 2005).

6.3 Aryabhatta Research Institute of Observational Sciences (ARIES)

6.3.1 Formation

The Aryabhatta Research Institute of Observational Sciences (ARIES) traces its origin to an astronomical observatory in Varanasi. It was established on April 20, 1954 at the initiative of Sampurnanand, the then education minister and later chief minister of Uttar Pradesh and A. N. Singh, a professor of mathematics at Lucknow University. The observatory was moved from the dust and haze of the plains to the more transparent skies of Nainital in 1955, and in 1961 to its present location at Manora Peak (longitude 79° 27′ 25.5″ E, latitude 29° 21′ 39.0″ N and altitude 1951 m), 4 km south of Nainital. It was known as the Uttar Pradesh State Observatory till the formation of Uttarakhand in November 2000, and as the State Observatory till its reincarnation on March 22, 2004 as an autonomous institute, with the name ARIES, under the Government of India.

6.3.2 Instruments

During 1955–60, the observatory was put on a sound footing by its director M. K. V. Bappu. The observatory had a gravity driven 25 cm, f/15 Cooke refractor telescope, acquired in 1955, a photoelectric photometer, that was designed and built at the observatory by Bappu. Photoelectric photometry of stars, comets and star clusters was initiated for the first time in India. Photographic observations of Mars, during its close approach in 1956, were made in yellow and red filters. The observatory acquired its first reflector, a 38 cm f/15 German mount in late 1959 and another reflector, 15 cm f/15 German mount in early 1960.

Beginning with the International Geophysical Year (1957–58), the observatory was the only centre in India (amongst 12 over the globe) for two decades, for optically tracking artificial satellites with a satellite tracking camera and a precision timing system capable of recording up to 10 millionth part of a second. The programme was undertaken in collaboration with the Smithsonian Astrophysical Observatory, USA with the aim to photograph the artificial earth satellites against the starry background so as to determine the shape and size of the earth, intercontinental distances, the nature and extent of the earth's atmosphere and of its magnetic field, and meteorological observations.

During 1960–82, many larger optical telescopes were acquired, support facilities were set up and back-end astronomical instruments developed indigenously at the institute. For stellar observations, the institute acquired a 52 cm reflector with Nasmyth and coudé foci mainly for spectrophotometry, and another 56 cm, f/15 reflector, with folded Cassegrain, for photoelectric photometry and spectrophotometry. For the latter, the fork mounting and the drive assembly were designed and built at ARIES and at a later stage when its primary mirror developed encrustations, the grinding and polishing of a new primary mirror was accomplished at the observatory. In 1972, a modern 1 m reflector (called the Sampurnanand telescope), was installed. The telescope has an equatorial English mounting with f/13 Cassegrain and f/31 coudé foci. Till the late 1980s, a Cassegrain plate holder, a Meinel camera, a photoelectric photometer, a near-infrared camera and a laboratory spectrum scanner were the main instruments. For solar observations, a 25 cm, f/66 off-axis skew Cassegrain telescope was acquired for spectral studies. It was equipped with an indigenously developed double pass

Fig. 6.18 Composite image of the nebular remnant of nova GK Persei through [O II], [O III] and [N II] emission lines

Fig. 6.19 Left: The 1 m Sampurnanand telescope at Manora Peak, Nainital. Right: A closer, full view of the telescope inside the dome

grating spectrograph, and a 45 cm coelostat. In the early 1990s, the institute acquired two 15 cm f/15 coudé refractors. These are being used to monitor active events on the surface of the sun.

During 1982–1996, a major augmentation of the astronomical facilities occurred through the acquisition of CCD-based imaging devices, computers, and new instruments for spectrophotometric observations. In 1989, a 384 × 576 pixel CCD was acquired. In the 1990s, larger CCDs were procured for observations with the 1 m telescope, which was upgraded to have a ST4 camera based auto-guider and a solid state drive system. For observations of solar active events, a fibre bonded CCD camera and another high speed camera for flares were acquired. In 2004, work on developing a new site commenced at Devasthal for the upcoming 1.3 m and 3.6 m optical telescopes, that are expected to be commissioned in late 2009 and 2013, respectively.

6.3.3 Research Highlights

Smaller telescopes were mostly used for stellar variability and galactic star cluster research and for studying eclipsing binaries. The first successful photoelectric observations, in the country, of the occultation of a star by a minor planet were done at the institute. The institute contributed to the detection of rings round Uranus, the detection of two additional rings around Saturn (one of these rings was detected for the first time) and to an independent discovery of rings around Neptune.

Fundamental star cluster parameters, viz. the age, distance, reddening and the mass function were determined for a sample of over 50 star clusters by observing their colour-magnitude diagrams. In one of the studies using reddening in the direction of star clusters, it was found that the plane defined by the interstellar dust is tilted with respect to the formal plane of the galaxy.

Solar astronomers at ARIES participated in five total solar eclipse events (including one in Antarctica) to investigate the structure of the solar chromosphere and corona. Using simultaneous high spatial (1.3 arc sec) and temporal (5 and 10 s) resolution, H_α observations from the 15 cm solar tower telescope at ARIES, multiple sausage oscillations in cool post-flare loops were detected. These observations have played an important role in understanding the dynamics of the lower solar atmosphere, complementing the reported observations of oscillations in the upper solar atmosphere (e.g. in hot flaring loops).

Using the 1 m Sampurnanand telescope, a rapidly oscillating Ap (roAp) star was discovered and three candidate roAp stars were identified. The roAp stars are amongst the most peculiar pulsating stars known so far where the pulsation modes are believed to be aligned with the strong, global magnetic fields, which is in contrast to the modes alignment with the rotation axis of a star in other pulsators. This unique property of the roAp stars allows a study of the effect of magnetic field on the chemical peculiarity and the pulsations.

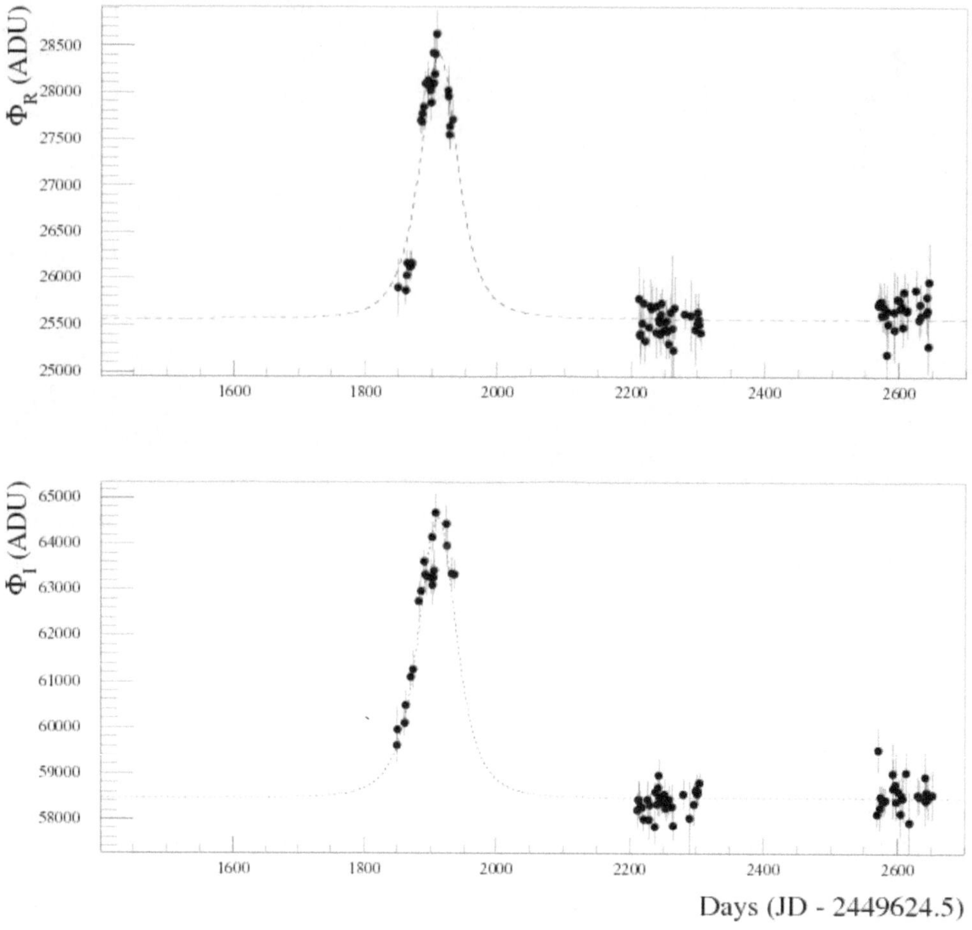

Fig. 6.20 The R and I band pixel light curves of the first microlensing detected from Nainital. The continuous line represents the result of the Paczynski fit

The roAp stars pulsate with periods between 5 and 21 minutes, with amplitudes from 0.6 milli mag to several milli mag. The detection of such pulsation requires excellent photometric sky conditions.

Observations of AGN using the 1 m Sampurnanand telescope, showed for the first time that the intra-night optical variability with an amplitude >3% was associated exclusively with (blazar) jet streams having optical polarization. This established a direct link between relativistic beaming and optical micro-variability. The AGN classes showing low (<3%) optical polarization exhibit very low level optical intra-night variability.

The Andromeda galaxy (M 31) has played an important role in understanding a wide variety of astrophysical problems. In recent years, it has been targeted to search for dark matter through microlensing of background galactic and (M 31) stars by their foreground massive compact objects. A 4-year long deep photometric survey, to the 21st magnitude in the R band, of a $13' \times 13'$ field towards (M 31), using the 1 m telescope at Nainital, resulted in the first successful detection of a microlensing event in the Andromeda galaxy (M 31) from an Indian observatory. The photometric analysis of the light curve showed that it reached a maximum of 20.1 mag and the (R-I) colour of the lensed star was estimated to be about 1.1 mag. The light curves indicate that the microlensing event was blended by red variable stars. Observations of optical afterglows of GRBs were started in 1996 and the first afterglow to be observed from India was at Nainital. A detailed investigation of the afterglow observations supports the existence of a non-isotropic emission from the sources and a collapsed star model for the progenitors.

6.4 Department of Astronomy, Osmania University

6.4.1 *Background*

The Department of Astronomy was formally established in 1959 by Osmania University, though research in astronomy at the university has a much older tradition going back to the Nizamiah Observatory. The university had already begun offering a course in astronomy since 1935 for undergraduate students, but it was not until 1960 that it was introduced at the postgraduate level as well.

Fig. 6.21 Left: The Japal-Rangapur Observatory. Right: A closer, full view of the 1.22 m telescope inside the dome.

As part of a programme to modernize astronomical research, the university acquired a 1.22 m reflector telescope and installed it at a new observatory in Japal-Rangapur, about 55 km from Hyderabad. In 1963, the University Grants Commission (UGC) recognized the Astronomy Department along with the Nizamiah and Japal-Rangapur observatories collectively as the Centre of Advanced Study in Astronomy (CASA). The latter status continued till 1979.

6.4.2 *Research Highlights*

The broad areas of research interest in Optical Astronomy at the Department of Astronomy, Osmania University are: (i) binary and variable stars, (ii) star clusters (iii) spectroscopic observations and (iv) galaxies.

6.4.2.1 Binary and Variable Stars

Many eclipsing binaries of Algol and RS CVn type were observed through UBV pass bands and the analysis of the light curves showed that the secondary components of the Algol type are not only over-luminous, but also hotter for their mass, indicating partial loss of their hydrogen envelopes. Waves in the light curves of the RS CVn binaries caused by spots were extensively studied. Several pulsating stars of Delta Scuti type were observed and their light curves analysed. Several B-emission objects were observed through UBVRI pass bands in order to understand the variable nature of these stars. Binary systems in which mass transfer phenomenon is a dominant process were also studied. Mass transfer in W UMa systems, with different mass ratios, was studied. Such an investigation provides information on the evolution of the contact binary system.

Three open clusters NGC 6791 (Rukmini et al. 2005), Berkeley 39 (Sriram et al. 2009) and NGC 7789, containing W UMa type variables, were studied in detail. Careful calculations were used to derive the various binary elements of W UMa type systems in the respective clusters. The mass ratio of the respective binary systems along with other binary elements were constrained. In one open cluster NGC 7789, two W UMa type binary systems were observed using the IUCAA 2 m telescope during 2006 and 2008. The phase light curve was investigated in detail. Since the number of studied binary systems in open clusters is low and statistics

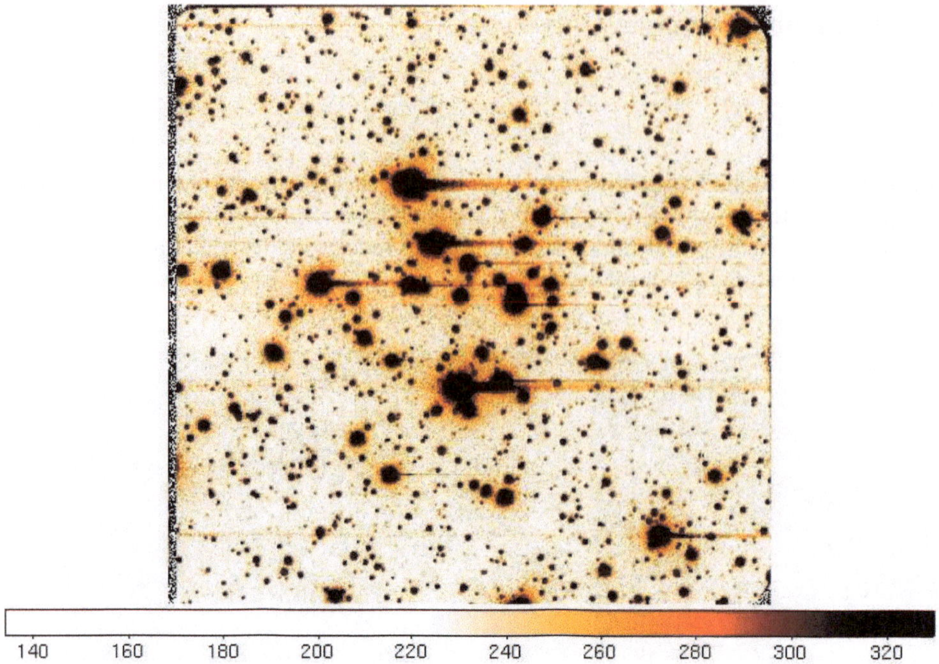

Fig. 6.22 B band image of NGC 189 with the HCT

obtained are not reliable, more studies are required to constrain the evolution of the respective binary systems and in turn to establish the age of the host cluster.

6.4.2.2 Spectroscopic Observations

A classification of Am stars on the basis of their MK spectral morphology was carried out. About 100 spectra of both Am and MK standards were obtained and their digital profiles were used for their classification. A phase modulated spectral line variation in some of these stars was detected. The pseudo-luminosity effects exhibited by these stars due to anomalous abundance of elements on their surface probably connected with surface magnetic field geometry are being examined for a better understanding of the atmospheric structure in these enigmatic stars. Also, a number of spectroscopic binaries were observed for determination of their orbital elements.

6.4.2.3 Star Clusters

Star clusters were studied using the 1.2 m telescope of the Physical Research Laboratory (PRL) at Mt. Abu (Hasan 2005a, 2005b; Hasan et al. 2008a). This investigation was complemented with photometric data from the 2MASS (2 micron all sky survey) database and observations from other national telescopes.

Various derived parameters like reddening, distance, age, mass functions as well as dynamical parameters like relaxation time and mass were used to study the structure as well as the dynamical state of clusters from a homogenous sample (2MASS) (Hasan et al. 2008b). Colour-magnitude diagrams and colour-colour synthetic diagrams were constructed to study the evolution of various clusters. The colour-magnitude diagrams are being examined closely for gaps as well as for other features. The variation in the mass function in different parts of the clusters has been analysed to study mass segregation and to test if it is due to dynamics or it is a property of the mass function.

6.4.2.4 Galaxies

Using Hubble Space Telescope images in four bands F435W, F606W, F775W and F850LP, the optical counterparts of the X-ray sources in the Chandra Deep Field South (CDFS) and in the Great Observatories Ori-

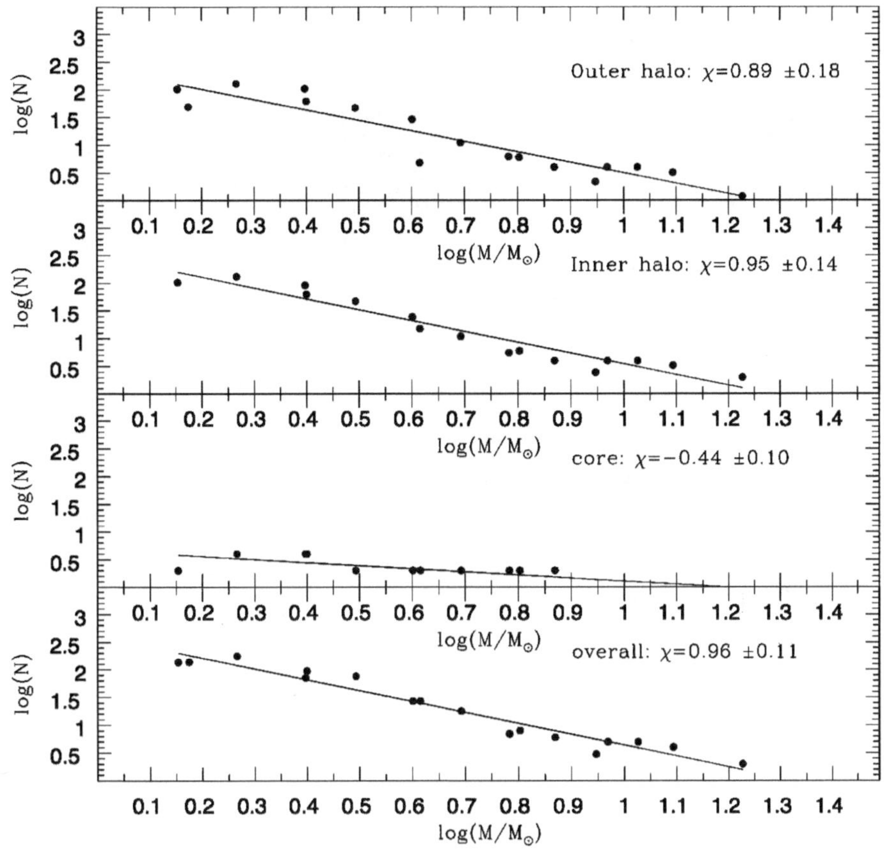

Fig. 6.23 Change in the mass function with radius

gins Deep Surveys (GOODS) South field were identified (Hasan 2007). A detailed study was made of these sources to study their morphological types. Decomposition of galaxy luminosity profiles, colour maps and visual inspection of 192 galaxies were identified as possible optical counterparts of Chandra X-ray sources in the CDFS-GOODS field. It was found that most moderate luminosity AGN hosts are bulge dominated in the redshift range ($z \approx 0.4 - 1.3$), but are not merging/interacting galaxies. This implies probable fueling of the moderate luminosity AGN by mechanisms other than those driven by merger.

6.5 Physical Research Laboratory

6.5.1 Background

Physical Research Laboratory (PRL) was founded on November 11, 1947 by Vikram Sarabhai with support from the Karmkshetra Educational Foundation and the Ahmedabad Education Society. The initial focus was research on cosmic rays and the properties of the upper atmosphere. PRL's contribution in optical astronomy began in the late 1970s. Before the Mt. Abu 1.2 m telescope (optimized for near-infrared work) was commissioned in 1994, PRL astronomers used other national telescope with their backend instruments for high-resolution spectroscopy, polarimetry and near-infrared photometry.

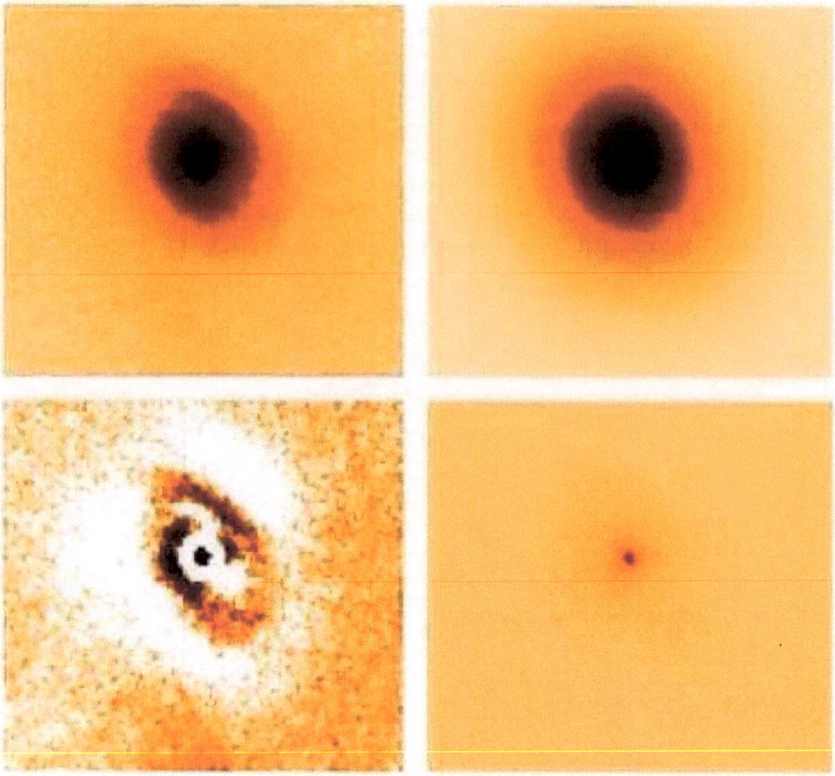

Fig. 6.24 Decomposition of the galaxy, model image, residual and point source

6.5.2 Research Highlights

6.5.2.1 First Integral Field Spectroscopy of Galactic PNe and H II regions

In India, PRL has been in the forefront of the use of specialized instruments for spectroscopy, such as the Fabry-Perot Interferometer (FPI). This instrument, built by PRL, has a high spectral resolution (a few times 10,000) and was used for observing solar coronal spectral line images and studying kinematics of planetary nebulae (PNe). Results on interferometric mapping of solar corona during total solar eclipses include the detection of large line widths attributed to macro-turbulence in the corona. PRL has also contributed towards building the country's first integral-field spectrometer utilizing the spatial multiplexing advantage of the FPI for studying the two-dimensional velocity fields in extended sources such as H II regions and PNe. Results from these efforts carried out at the Mt. Abu Observatory include the detection of a possible, rare quadrupolar PN that paved the way for three dimensional morphological modeling of PNe in order to understand their evolution and shape.

6.5.2.2 First Polarimetry of Evolved Stars and Comets

During the 1980s, PRL built the country's first state-of-the-art optical polarimeter, in collaboration with the University of Arizona, USA, for detailed studies of comets, asymptotic giant branch (AGB) and T Tauri stars, dark clouds, and AGN. Observations of some peculiar stars, such as R Aquarii, led to the detection of a precessing jet. Dust characteristics and its evolution have also been investigated in several comets.

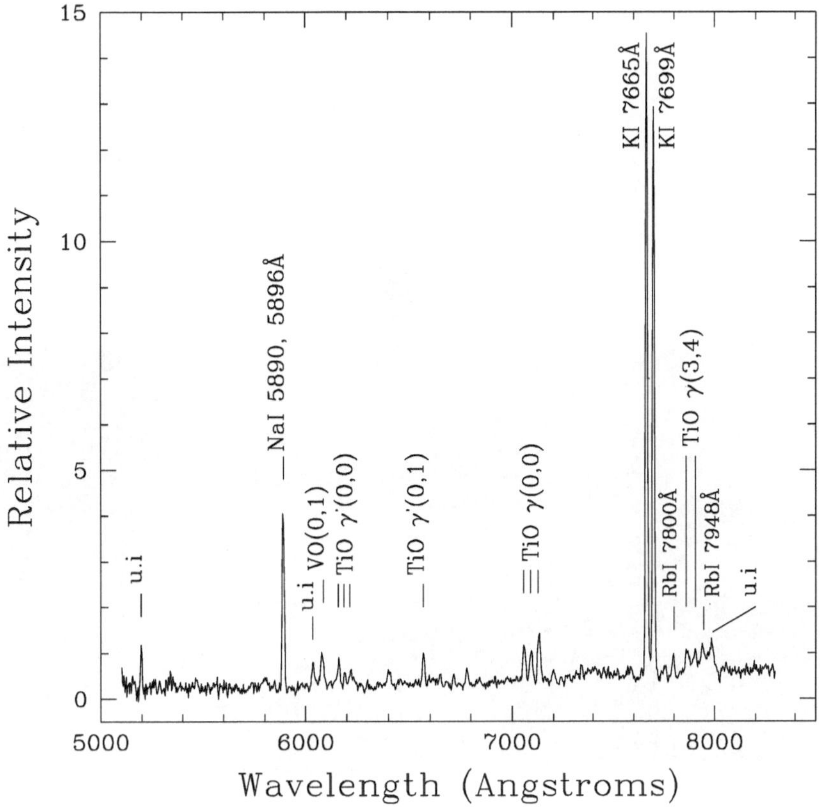

Fig. 6.25 The optical spectrum of V4332 Sgr from the HCT

6.5.2.3 A New Class of Eruptive Variable

The unusual variable star V4332 Sgr underwent a mysterious outburst in 1994 to become a cool red giant star of spectral type M8-9. Its striking similarity to the remarkable variable star V838 Mon that erupted in 2002 triggered the interest of PRL scientists to study the evolution of this new class of eruptive variables. The optical spectrum of V4332 Sgr displays exceptionally strong emission lines of alkali metals like potassium and sodium, further supporting the idea that it belongs to a new class of an eruptive variable star.

The PRL group is currently engaged in building a specialized, high precision and high resolution (3–5 m/s in velocity) echelle spectrograph to detect extra-solar planetary systems around nearby stars. It is likely to become operational towards the end of 2009.

6.5.3 Near-Infrared Astronomy

With the aim of initiating investigations on star formation and late stages of stellar evolution, PRL built the country's first near-infrared 1.2 m telescope at Mt. Abu. The telescope began operations in November of 1994, and has since then been equipped with several specialized instruments. These include an integral field Fabry-Perot spectrometer for imaging both in the visible and near-infrared regions, an optical polarimeter, a fast-readout near-infrared photometer (later upgraded to include two channels for simultaneous photometry) for lunar occultation and a near-infrared camera and spectrograph (NICMOS). Recently the laboratory has acquired and assembled a new wide-field ($8' \times 8'$) near-infrared camera and spectrometer with a 1K × 1K detector array.

Some of the significant results from Mt. Abu include, (i) first detection of a possible helium nova (V445 Puppis); (ii) detection of AlO molecular lines and light echo from the peculiar Nova V838 Monocerotis; (iii) alumina dust as the possible first condensate in the ejecta of some novae; (iv) detection of asymmetry in the dust envelope around a carbon star IRC 10216 using lunar occultation technique; (v) the detection of

Fig. 6.26 The 1.2 m PRL telescope at Mt. Abu

thermally driven molecular hydrogen out-flows in a low mass proto-star RNO9 and also in the intermediate mass star forming region (IRAS 06061+2151) indicating the possibility of formation of massive stars by accretion just as the low mass counterparts; and (vi) dust features in starburst galaxies. Some of these results also included observations made from other observatories as well as space-based platforms.

6.5.4 Udaipur Solar Observatory (USO)

Established in 1975 and taken over by ISRO during 1981 as a part of PRL, USO has contributed to solar optical observational data on a regular basis by obtaining full disk H_α images using a 6 ft Razdow telescope acquired from the Space Environment Service Center of NOAA (National Oceanic and Atmospheric Administration), Boulder, USA in 1987. A video magnetograph, designed and fabricated by USO, was added in 1999 for obtaining the line-of-sight component of the photospheric magnetic fields. In order to measure the transverse component of magnetic field, a vector magnetograph (VM) was built in 2006. The VM of USO fills up the existing gap between Europe and China and is one the very few such instruments available today. In 1995 USO became a part of the six-station network called the Global Oscillations Network Group (GONG), thus becoming the only station in India to have been chosen for the purpose of global continuous coverage for probing the solar interior. The GONG programme has helped in unravelling some of the most fundamental problems in solar (and stellar) astrophysics.

Currently USO is engaged in building a solar telescope of 50 cm aperture with a host of modern diagnostic equipments such as precision spectroscopy, polarimetry on solar photosphere and chromosphere at high spatial and temporal resolutions. The telescope is expected to be commissioned by 2011.

Fig. 6.27 The 2 m telescope at the IUCAA Girawali Observatory

6.6 Inter-University Centre for Astronomy and Astrophysics (IUCAA)

6.6.1 Background

In the mid 1980s, a new idea of creating inter-university centres for thrust science and technology areas emerged amidst concerns of a considerable decline in interest among bright students to take up pursuits in scientific research. Amongst a number of mandates these centres would have, two important ones were to: (a) make world class facilities accessible to researchers in the university sector, and (b) nucleate, nurture and develop research and curricula in universities and colleges. Astronomy was one of the areas to gain from these efforts, largely through the efforts of Yash Pal, J. Narlikar and others, that culminated in the setting up of the IUCAA at Pune in 1988. In the current era in which multiwavelength studies have become the paradigm for research in astronomy, IUCAA's proximity to the National Centre for Radio Astrophysics (NCRA) within the Pune University campus enables it to exploit the synergy between radio and optical astronomy. IUCAA has a small core academic team of about 20 faculty members. At present, about a third or more of this group has active interests in the broad class of optical astronomy - both observational as well as instrumentation development. Beyond the work of the core faculty, the bulk of IUCAA's research productivity results from its active and vigorous associateship programme. Through this programme, currently about seventy faculty members from Indian universities and postgraduate colleges visit and make use of IUCAA facilities for their research activities.

6.6.2 Instrumentation Programme

With electronics and optics laboratories housing dark rooms, clean rooms, design centres and integration facilities, IUCAA's instrumentation programme resulted in the development of a number of instruments like polarimeters, spectrographs, focal plane array (CCD, CMOS) controllers and data acquisitions systems, backend instrument controllers, etc. The instrumentation programme has also been active in developing both amateur (e.g. night sky photometers) as well as university education (automated photoelectric telescopes, low-cost CCD cameras) class instruments. After the first decade of its life which was mainly spent on consolidation, IUCAA's subsequent efforts turned towards expansion of its facilities. This resulted in the setting up of the IUCAA Girawali Observatory (IGO) in 2006, which houses a modern 2 m optical telescope.

6.6.3 Research Highlights

6.6.3.1 Stellar Spectral Library

A major programme has recently been completed that involves the compilation of a digital stellar spectral library of about 1273 stars as a collaboration with the University of Delhi, the University of Carolina, Chapel Hill, and the National Optical Astronomy Observatory (NOAO), Tucson. The stars were classified into 7 broad types (O, B, A, F, G, K and M) with the O-type being the hottest and M-type being the coolest.

All observations for the spectral library were obtained with the coudé feed 0.9 m telescope at the Kitt Peak National Observatory of NOAO. The coudé feed is an auxiliary telescope that feeds a stellar image onto the slit of the coudé spectrograph of the main 2.1 m telescope. Two spectrograph combinations were optimized for the red and blue portions of the spectra. The entire spectral region 340–950 nm was covered with five grating settings.

A near infrared (NIR) spectral library has been completed and released comprising the J-Band (126 stars), H-Band (135 stars) and K-Band (114 stars) of solar type stars covering spectral types O5–M3 and luminosity classes I–V as per the MK classification. The observations were carried out during 2000–2004 with the 1.2 m PRL infrared telescope using a the near infrared array based spectrometer.

The spectra have a moderate resolution of 1000 (at 1.6 nm) in the J, H and K Bands and have been continuum shape corrected to their respective effective temperatures. This library will serve as an important database for stellar population synthesis and other applications in conjunction with the newly formed large optical coudé feed spectral library described above.

6.6.3.2 Temperature of the Cosmic Microwave Background (CMB)

The first measurement of the temperature of the cosmic microwave background, T_{CMBR}, at high redshift z was carried out using the QSOs spectrum obtained with the VLT (very large telescope) (Srianand et al. 2000). This measurement was obtained in a unique absorption system where absorption lines of neutral carbon in the three fine-structure levels of the ground term are observed together with absorption lines of singly ionized carbon in its excited fine-structure level, and absorption lines of molecular hydrogen in the various rotational levels. The population and depopulation of the first excited rotational level of molecular hydrogen from and to the ground state is controlled by thermal collisions. Therefore, the excitation temperature is approximately equal to the kinetic temperature. The fine structure upper level of the carbon ground-state triplet is mostly populated by collisions and depopulated by radiative decay. Therefore, once the temperature is known, the particle density can be derived from the carbon line ratio. Finally, the UV radiation flux can be constrained from the populations of the higher rotational levels of molecular hydrogen. Thus, it is now possible to constrain, directly from observations, the different excitation processes at play and determine a lower limit on the temperature of the CMB black-body spectrum (Srianand et al. 2000).

This technique has been extended to a large number of systems over a wide redshift range to get the redshift evolution of the temperature of CMB radiation. All the results till now are consistent with the standard Big Bang cosmology (Srianand et al. 2005; Ledoux, Petitjean and Srianand 2006). Recently, Srianand et al. (2008) reported the first ever detection of CO (carbon monoxide) in damped Lyman alpha (DLA) systems. Using its rotational level population they made the most precise measurement till date of T_{CMBR} at $z = 2.4$.

6.6.3.3 Time Variation of Fundamental Constants

The time and space variation of fine-structure constant has been investigated using a detailed many multiplet analysis. Using high spectral resolution observations of 23 Mg II systems towards, 18 QSOs in the redshift range $0.4 \leq z \leq 2.3$ were detected with the VLT (Srianand et al. 2004; Chand et al. 2004) and a strong constraint was obtained on the variations of the fundamental constants.

6.6.3.4 Molecular Hydrogen at High Red Shift *z*

The neutral hydrogen molecule is the first neutral molecule to be formed in the universe and is an important coolant for the first generation of stars. Not much was known about this molecule outside the local universe. A systematic survey of this molecule using the VLT has been recently carried out (Noterdaeme et al. 2008; Srianand et al. 2005). The survey has resulted in a large data base that has been been used to: (a) understand the physical conditions in the protogalaxies that can not be detected in their own light (b) investigate the nature of dust, out-flows and rotational velocity of proto-galaxies at high redshifts, and (c) constrain the time and space variation of electron to proton mass ratio.

6.6.3.5 21 cm Absorption Systems at High *z*

Similar to molecular hydrogen, detecting the 21 cm absorption line at high redshift z will enable one to probe the physical conditions of the gas and the time and space variations of fundamental constants like α, μ and proton G factor. Recently, a systematic survey has been started of 21 cm absorption lines using the Giant Meterwave Radio Telescope (GMRT). The survey is just completed and has resulted in 9 firm detections of 21 cm absorption systems. This, already, has more than doubled the number of 21 cm absorption systems at $z > 1.0$ (Gupta and Srianand 2007; Gupta et al. 2009).

Acknowledgements The section on 'Historical Perspective' draws heavily on unpublished material from the IIA archives. I am grateful to Drs. B. G. Anandarao, G. C. Anupama, C. Birdie, H. C. Bhatt, N. Hasan, B. Kumar, D. C. V. Mallik, A. N. Ramprakash and A. Subramaniam for providing inputs that went into this chapter. It is a pleasure to acknowledge the extensive support of Dr. D. C. V. Mallik, particularly for carefully going through the manuscript and making valuable suggestions.

References

1. Anupama GC, Prabhu TP (1991) MNRAS 253:605
2. Anupama GC, Prabhu TP (1993) MNRAS 263:335
3. Anupama GC, Kembhavi AK, Prabhu TP. et al. (1994) A&AS 103:315
4. Anupama GC, Sivarani T, Pandey G(2001) A&A 367:506
5. Anupama GC, Sahu DK, Jose J (2005a) A&A 429:667
6. Anupama GC, Sahu DK, Deng J. et al.(2005b) ApJ 631:L125
7. Anupama GC, Kantharia NG (2005) A&A 435:167
8. Anupama GC, Kamath US, Muneer S et al. (2008) In: Proc. 25th Meeting of the Astron. Soc. India, ed. K. Subramaniam and D. J. Saikia, p. 60
9. Anupama GC, Sahu DK, Gurugubelli UK, et al. (2009) MNRAS, 392:894
10. Banerjee DPK, Ashok NM (2004) ApJ 604:L57
11. Bappu MKV, Mekkaden MV, Rao NK, et al. (1984) BASI 12:196

12. Bhatt HC (1994) IAUC 6039
13. Bhatt HC, Maheswar G, Manoj P (2004) MNRAS 348:83
14. Bhattacharyya JC, Bappu MKV (1977) Nature 270: 503
15. Bhattacharyya JC,Kuppuswamy K (1977) Nature 267:331
16. Bhavya B, Mathew B, Subramaniam A (2007) BASI 35: 383
17. Carlson RW, Bhattacharyya JC, Smith BA, et al. (1973) Science 182:53
18. Carraro A, Subramaniam A, Janes KA (2006) MNRAS 371:1301
19. Chand H, Srianand R, Petitjean P, et al. (2004) A & A, 417:853
20. Chauhan N, Pandey AK, Ogura K, et al. (2009) MNRAS 396:964
21. Cowsik R, Prabhu TP, Anupama GC, et al. (2001) BASI 29:157
22. Das M, Kantharia N, Ramya, et al. (2007) MNRAS 39:11
23. Ferro AA, Arevalo MJ, Lazaro C, et al. (2004) RMxAA 40:209
24. Ferro AA, Rojas Lpez V, Giridhar S, et al. (2008) MNRAS 384:1444
25. Ganesh KS and Bappu MKV (1967) Kodaikanal Obs. Bull. 183:77
26. Giridhar S (1983) JAA 4:75
27. Giridhar S, Rao NK, Lambert DL (1994) ApJ 437:476
28. Gopal-Krishna, Gupta AC, Sagar R, et al. (2000) MNRAS 314:815
29. Goswami A (2005) MNRAS 359:531
30. Goswami A, Aoki M, Beers TC, et al. (2006) MNRAS 372:343
31. Goswami A, Bama P, Shantikumar NS, et al. (2007) BASI 35:339
32. Goto T, Ojha DK (2006) BASI 34:291
33. Goyal A, Gopal-Krishna, Sagar R, et al. (2007) BASI 35:141
34. Gupta N, Srianand R, Petitjean P, et al. (2007) ApJ 654:L111
35. Gupta N, Srianand R, Petitjean P, et al. (2009) arXiv/0904.2878
36. Hasan P (2005a) BASI 33:151
37. Hasan P (2005b) BASI 33: 311
38. Hasan P (2007) ApSS 312:63
39. Hasan P, Kilambi GC, Hasan SN (2008a) ApSS 313:363
40. Hasan P, Hasan SN, Shah U (2008b) ApSS 318:25
41. Hric L, Petrik K, Urban Z, et al. (1998)A&A 339:449
42. Kaur R, Paul B, Kumar B, et al. (2008) MNRAS 386:2253
43. Kochhar R, Narlikar J (1995) Astronomy in India: A Perspective, New Delhi: Indian National Science Academy
44. Lambert DL and Rao NK (1994) JAA 15:47
45. Lambert DL and Rao NK (1996) ASP Conf. Ser 96:43
46. Ledoux C, Srianand R, Pelitjean P(2006) ApJ 640:L25
47. Maiti M, Sengupta S, Parihar PS, et al. (2005) ApJ, 619, L183
48. Maiti M (2007) AJ, 133:1633
49. Mallik DCV, Sagar R, & Pati AK (1995) A& AS 114:537
50. Mallik SV (1982) JAA 3, 391
51. Mallik SV (1993) ApJ 402:303
52. Mallik SV (1994) A& AS 124:359
53. Mallik SV, Parthasarathy M,Pati AK (2003) A& A 409:251
54. Manoj P, Bhatt HC, Maheswar G, et al. (200 6) ApJ 653:657
55. Mathew B, Subramaniam A, Bhatt BC (2008) MNRAS 388:1879
56. Mayya YD (1995) AJ 109:2503
57. Meech KJ, Ageorges N, A'Hearn MF, et al. (2005) Sci. 310:265
58. Misra K, Sahu DK, Anupama GC, et al. (2008) MNRAS 389:706
59. Mekkaden MV (1998) A&A 340:135
60. Mekkaden MV (1999) A&A 344:111
61. Mohin S, Raveendran AV (1987) JAA 8:389
62. Mohin S, Raveendran AV (1989) JAA 10:35
63. Mohin S, Raveendran AV (1992) A& A 256:487
64. Mohin S, Raveendran AV (1993a) A& A 277:155
65. Mohin S, Raveendran AV (1993b) A& AS 100:331
66. Mohin S, Raveendran AV (1994) A& A 286:824
67. Mohin S, Raveendran AV (1995) A& A 301:788
68. Ogura K, Chauhan N, Pandey AK, et al. (2007) PASJ 59:199
69. Ojha DK,Ghosh SK, Tej A, et al. (2006) MNRAS 368:825
70. Pandey G (2006) ApJ 648L:L143
71. Pandey G, Lambert DL, Rao NK, et al. (2004) ApJ 602:L113
72. Pandey G (2006) ApJ 648:L143
73. Pandey AK, Sharma S, Ogura K, et al. (2008) MNRAS 383:1241
74. Pandey SB, Anupama GC, Sagar R, et al. (2003a) MNRAS 340:375
75. Pandey SB, DK Sahu, DK, Resmi L, et al. (2003b) BASI 31:19
76. Pandey SB, Anupama GC, Sagar R, et al. (2003c) A& A 408:21
77. Pandey SB, Sagar R, Anupama GC, et al. (2004) A& A 417:919
78. Parihar PS, Sahu DK, Bhatt BC, Subramaniam A, et al. (2003) BASI 31:453
79. Parihar PS, Messina S, Bama P, et al. (2009) MNRAS (in press)
80. Parthasarathy M, Raveendran AV, Mekkaden MV (1981) ApSS 74:87

81. Parthasarathy M, Jain SK, Bhatt HC (2000) A& A 355:221
82. Patil MK, Pandey SK, Sahu, et al. (2007) A&A 461:103
83. Prabhu TP (1980) JAA 1:129
84. Prabhu TP, Mayya YD, Singh KP, et al. (1995) A& A 295:403
85. Raiteri CM, Villata M, Ibrahimov MA, et al. (2005) A& A 438:39
86. Rajamohan R (1976) Pramana 37:160
87. Ramya S, Sahu DK, Prabhu TP (2007) MNRAS 381:511
88. Ramya S, Sahu DK, Prabhu TP (2009) MNRAS (in press)
89. Raveendran AV, Mekkaden MV, Mohin S (1982) MNRAS 199:707
90. Raveendran AV (1999) MNRAS 303:595
91. Raveendran AV, Mohin S (1995) A&A 301:788
92. Resmi L, Ishwara-Chandra CH, Castro-Tirado AJ, et. al. (2005) A&A 440:477
93. Rao NK, Vagiswari A, Louis C (1984) BASI 12:81
94. Rukmini J, Vivekananda Rao P, Sriram K (2005) ApSS 299:109
95. Sahu DK, Anupama GC, Prabhu TP (2006) MNRAS 366:682
96. Sahu DK, Tanaka M, Anupama GC, et al. (2008) ApJ 680:580
97. Sahu DK, Tanaka M, Anupama GC, et al. (2009) ApJ 697:676
98. Scaria KK (1979) Ph.D. thesis, Madurai Kamaraj University
99. Scaria KK, Bappu MKV (1981) JAA 2:215
100. Sharma S, Ogura K, Pandey AK, et al. (2007) BASI 35:53
101. Solheim JE, Provencal JL Bradley PA, et al. (1998) A&A 332:939
102. Srianand R, Petitjean P & Ledoux C (2000) Nature 408:931
103. Srianand R, Petitjean P & Ledoux C, et al. (2005) MNRAS 362:549
104. Srianand R, Noterdaeme P, Ledoux C, et al. (2008) A& A 482:L39
105. Srianand R, Chand H, Petitjean P, et al. (2004) PRL 92:121302
106. Sriram KY, Kiron R, Vivekananda Rao, P. (2009) RAA (in press)
107. Stalin CS, Hegde M, Sahu DK, et al. (2008) BASI 36:111
108. Stirpe GM, Winge C, Altieri B, et al. (1994) ApJ 425:609
109. Subramaniam A, Sagar R (1999) AJ 117:937
110. Sujatha S, Babu GSD (2006) ApSS 305:399
111. Sujatha S, Babu GSD, Ananthamurthy S (2006) BASI 34:357
112. Subramaniam A, Sahu DK, Sagar R, et al. (2005) A&A 440:511
113. Subramaniam A, Mathew B, Bhatt BC, et al. (2006) MNRAS 370:743
114. Surendiranath R, Rao NK, Sagar R, et al. (1990) JAA 11:151
115. Szkody P,Mukadam A, Gánsicke BT, et al. (2007) ApJ 658:1188
116. Tominaga N, Tanaka M, Nomoto K, et al. (2005) ApJ 633:L97
117. Vasundhara R (1994) A&A 281:565
118. Vasundhara R, Chakraborty P (1999) Icarus 140:221
119. Vasundhara R, Chakraborty P (2004) ApJ 616:1278

Colored Figures

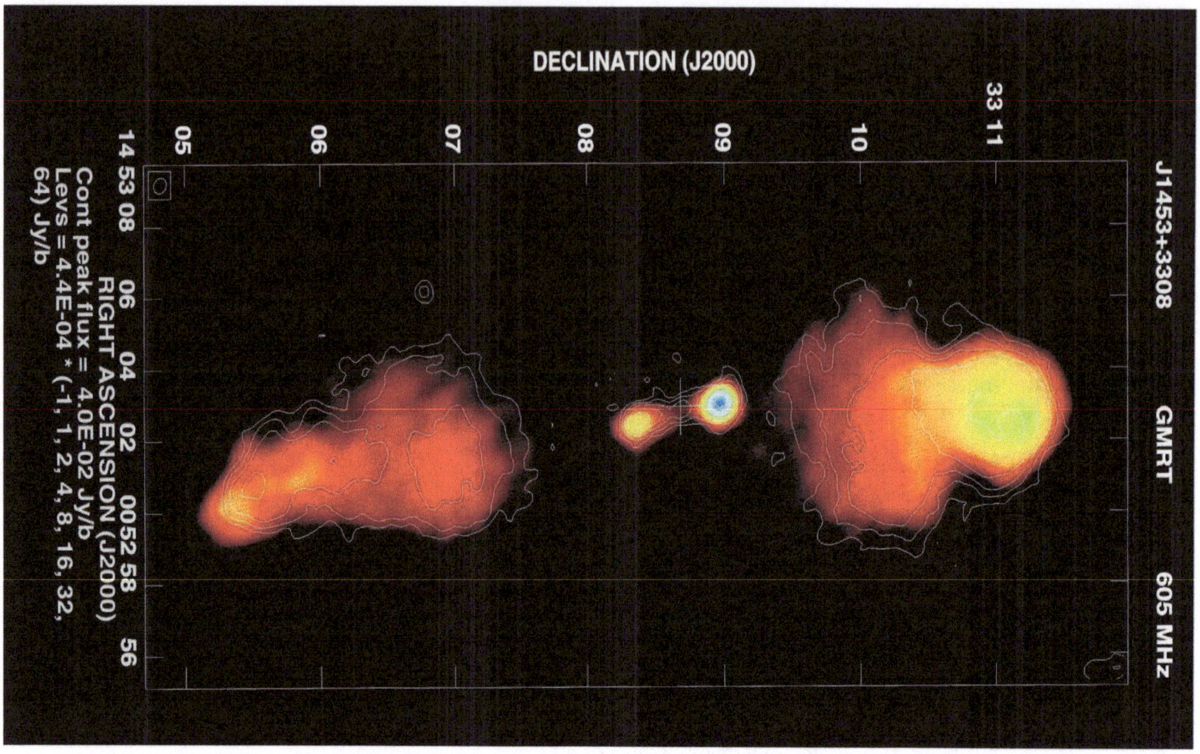

Govind Swarup: Figure 5.17, Page no 90

Govind Swarup: Figure 5.18, Page no 90

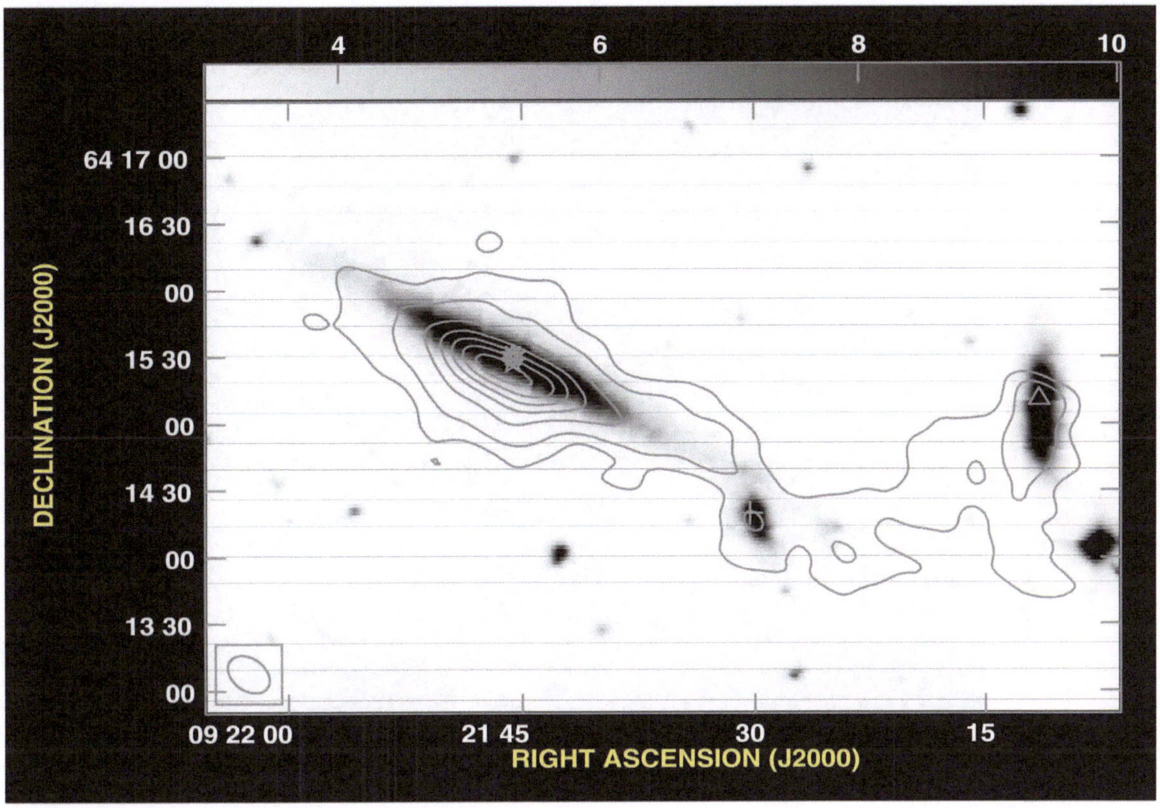

Govind Swarup: Figure 5.19, Page no 91

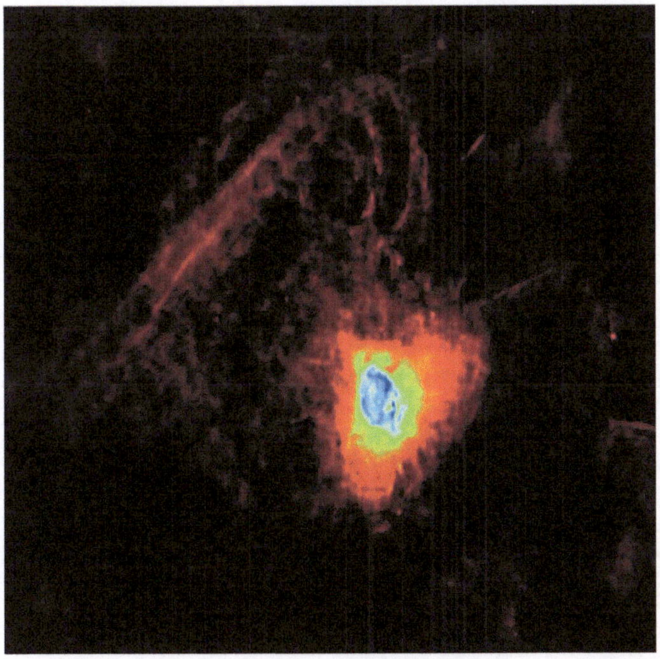

Govind Swarup: Figure 5.20, Page no 92

Govind Swarup: Figure 5.23, Page no 96

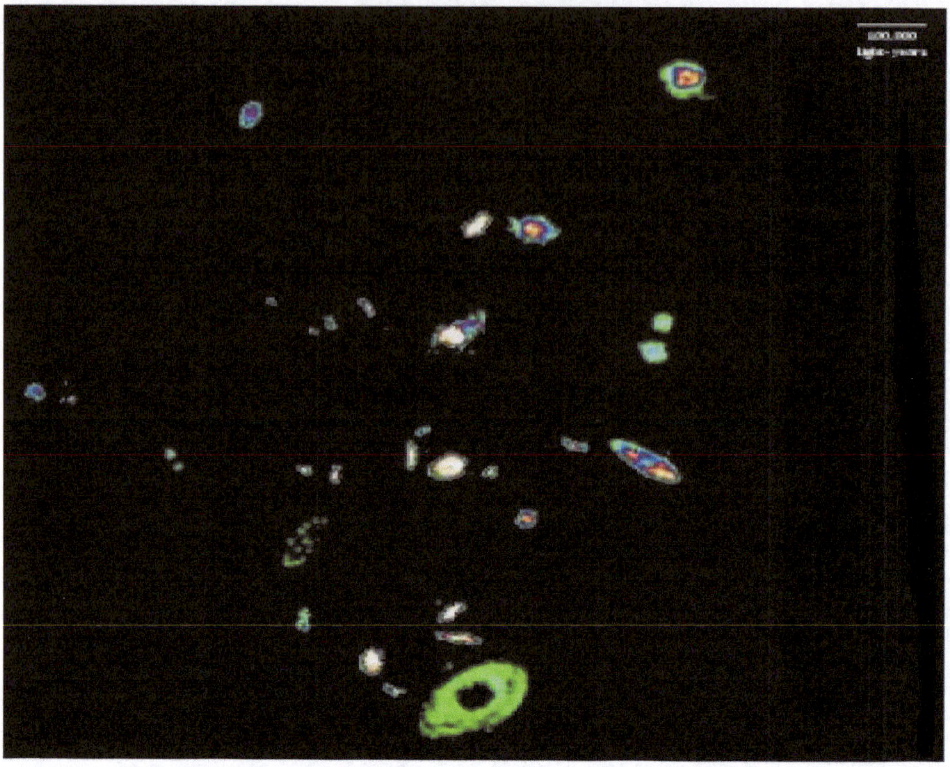

Govind Swarup: Figure 5.30, Page no 103

Govind Swarup: Figure 5.31, Page no 104

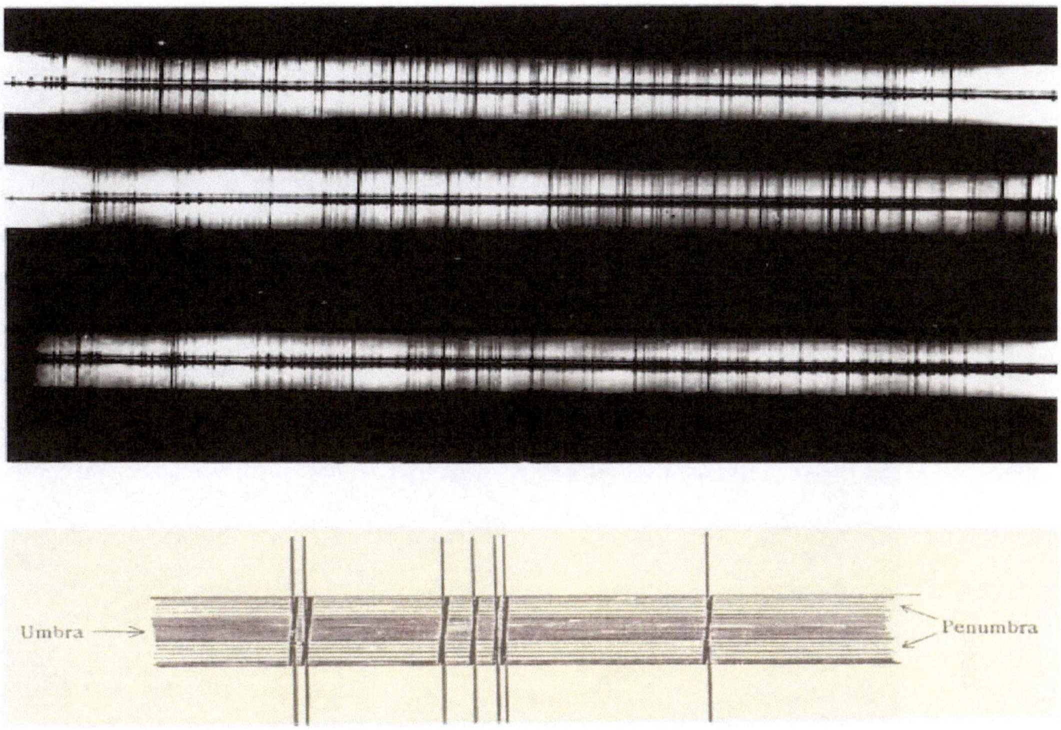

Hasan: Figure 6.2, Page 113

Hasan: Figure 6.3, Page 116

Hasan: Figure 6.5, Page 118

Hasan: Figure 6.5, Page 118

Hasan: Figure 6.11, Page 126

Hasan: Figure 6.12, Page 126

Hasan: Figure 6.19, Page 133

Hasan: Figure 6.19, Page 133

Hasan: Figure 6.21, Page 135

Hasan: Figure 6.21, Page 135